Mathematical Olympiad in China (2009–2010)

Problems and Solutions

Mathematical Olympiad Series

ISSN: 1793-8570

Published

Vol. 1 A First Step to Mathematical Olympiad Problems
by Derek Holton (University of Otago, New Zealand)

Vol. 2 Problems of Number Theory in Mathematical Competitions
by Yu Hong-Bing (Suzhou University, China)
translated by Lin Lei (East China Normal University, China)

Vol. 3 Graph Theory
by Xiong Bin (East China Normal University, China) &
Zheng Zhongyi (High School Attached to Fudan University, China)
translated by Liu Ruifang, Zhai Mingqing & Lin Yuanqing
(East China Normal University, China)

Vol. 4 Combinatorial Problems in Mathematical Competitions
by Yao Zhang (Hunan Normal University, P. R. China)

Vol. 5 Selected Problems of the Vietnamese Olympiad (1962–2009)
by Le Hai Chau (Ministry of Education and Training, Vietnam)
& Le Hai Khoi (Nanyang Technology University, Singapore)

Vol. 6 Lecture Notes on Mathematical Olympiad Courses:
For Junior Section (In 2 Volumes)
by Xu Jiagu

Vol. 7 A Second Step to Mathematical Olympiad Problems
by Derek Holton (University of Otago, New Zealand &
University of Melbourne, Australia)

Vol. 8 Lecture Notes on Mathematical Olympiad Courses:
For Senior Section (In 2 Volumes)
by Xu Jiagu

Vol. 9 Mathemaitcal Olympiad in China (2009–2010)
edited by Bin Xiong (East China Normal University, China) &
Peng Yee Lee (Nanyang Technological University, Singapore)

Mathematical Olympiad in China (2009–2010)

Problems and Solutions

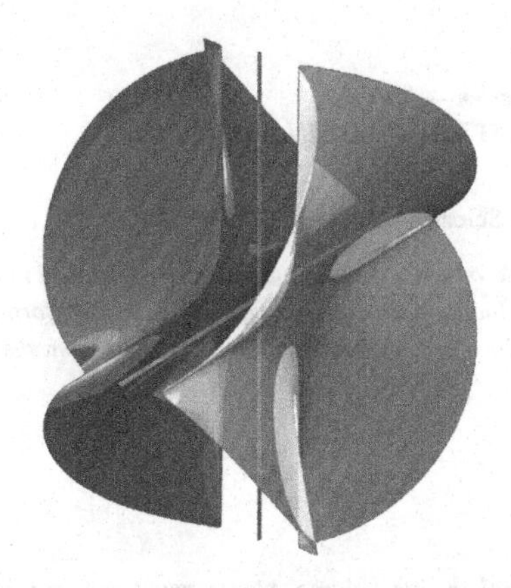

Editors

Xiong Bin

East China Normal University, China

Lee Peng Yee

Nanyang Technological University, Singapore

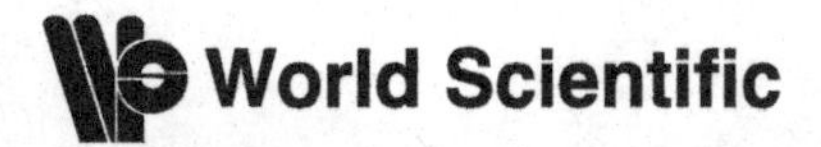

Published by

World Scientific Publishing Co. Pte. Ltd.
5 Toh Tuck Link, Singapore 596224
USA office: 27 Warren Street, Suite 401-402, Hackensack, NJ 07601
UK office: 57 Shelton Street, Covent Garden, London WC2H 9HE

British Library Cataloguing-in-Publication Data
A catalogue record for this book is available from the British Library.

Mathematical Olympiad Series — Vol. 9
MATHEMATICAL OLYMPIAD IN CHINA (2009–2010)
Problems and Solutions

ISBN 978-981-4390-21-7 (pbk)

Printed in Singapore by World Scientific Printers.

Editors

XIONG Bin *East China Normal University, China*

Lee Peng Yee *Nanyang Technological University*

Original Authors

MO Chinese National Coaches of 2009 - 2010

English Translators

XIONG Bin *East China Normal University, China*

FENG Zhigang *Shanghai High School, China*

WANG Shanping *East China Normal University, China*

Yao Yijun *Fudan University, China*

Qu Zhenhua *East China Normal University, China*

Copy Editors

NI Ming *East China Normal University press, China*

ZHANG Ji *World Scientific Publishing Co., Singapore*

WONG Fook Sung *Temasek Polytechnic, Singapore*

KONG Lingzhi *East China Normal University press, China*

Preface

The first time China participate in IMO was in 1985, two students were sent to the 26th IMO. Since 1986, China has a team of 6 students at every IMO except in 1998 when it was held in Taiwan. So far, up to 2011, China has achieved the number one ranking in team effort 17 times. A great majority of students received gold medals. The fact that China obtained such encouraging result is due to, on one hand, Chinese students' hard working and perseverance, and on the other hand, the effort of teachers in schools and the training offered by national coaches. As we believe, it is also a result of the education system in China, in particular, the emphasis on training of basic skills in science education.

The materials of this book come from a series of two books (in Chinese) on *Forward to* IMO: *a collection of mathematical Olympiad problems* (2009 – 2010). It is a collection of problems and solutions of the major mathematical competitions in China. It provides a glimpse of how the China national team is selected and formed. First, there is the China Mathematical Competition, a national event. It is held on the second Sunday of October every year. Through the competition, about 200 students are selected to join the China Mathematical Olympiad (commonly known as the winter camp), or in short CMO, in the next January. CMO lasts for five days. Both the type and the difficulty of the problems match those of IMO. Similarly, students are given three problems to solve in 4.5 hours each day. From CMO, about 50 to 60 students are selected to form a national training team. The training takes place for two weeks in the month of March. After four to six tests, plus two qualifying examinations, six students are finally selected to form the national team, taking part in IMO in July of that year.

In view of the differences in education, culture and economy of western part of China in comparison with the coast part in east China, mathematical competitions in West China did not develop as fast as in the rest of the country. In order to promote the activity of mathematical competition, and to enhence the level of mathematical competition, starting from 2001, China Mathematical Olympiad Committee organizes the China Western Mathematical Olympiad. The top two winners will be admitted to the national training team. Through the CWMO, there have been two students entering the national team and receiving gold medals for their performance at IMO.

Since 1995, for a quite long period there was no female student in the Chinese national team. In order to encourage more female students participating in the mathematical competition, starting from 2002, China Mathematical Olympiad Committee conducted the China Girls' Mathematical Olympiad. Again, the top two winners will be admitted directly into the national training team. In 2007, the first girl who was winner of China Girls' Mathematical Olympiad was selected to enter the 2008 China national team and won the gold medal of the 49th IMO.

The authors of this book are coaches of the China national team. They are Xiong Bin, Li Shenghong, Leng Gangsong, Wu Jianping, Chen Yonggao, Li Weigu, Yu Hongbing, Zhu Huawei, Feng Zhigang, Liu Shixiong, Qu Zhenhua, Wang Weiye, and Zhang Sihui. Those who took part in the translation work are Xiong Bin, Feng Zhigang, Wang Shanping, Yao Yijun, and Qu Zhenhua. We are grateful to Qiu Zonghu, Wang Jie, Wu Jianping, and Pan Chengbiao for their guidance and assistance to authors. We are grateful to Ni Ming of East China Normal University Press. Their effort has helped make our job easier. We are also grateful to He Yue of World Scientific Publishing for her hard work leading to the final publication of the book.

Authors
October 2011

Introduction

Early days

The International Mathematical Olympiad (IMO), founded in 1959, is one of the most competitive and highly intellectual activities in the world for high school students.

Even before IMO, there were already many countries which had mathematics competition. They were mainly the countries in Eastern Europe and in Asia. In addition to the popularization of mathematics and the convergence in educational systems among different countries, the success of mathematical competitions at the national level provided a foundation for the setting-up of IMO. The countries that asserted great influence are Hungary, the former Soviet Union and the United States. Here is a brief review of the IMO and mathematical competition in China.

In 1894, the Department of Education in Hungary passed a motion and decided to conduct a mathematical competition for the secondary schools. The well-known scientist, *J. von Etövös*, was the Minister of Education at that time. His support in the event had made it a success and thus it was well publicized. In addition, the success of his son, *R. von Etövös*, who was also a physicist, in proving the principle of equivalence of the general theory of relativity by *A. Einstein* through experiment, had brought Hungary to the world stage in science. Thereafter, the prize for mathematics competition in Hungary was named "*Etövös* prize". This was the first formally organized mathematical competition in the world. In what follows, Hungary had indeed produced a lot of well-known scientists including *L. Fejér*, *G. Szegö*, *T. Radó*, *A. Haar* and *M. Riesz* (in real analysis), *D. König* (in combinatorics), *T. von Kármán* (in aerodynamics), and *J. C. Harsanyi* (in game theory), who had also won the Nobel Prize for Economics in 1994. They all were the winners of Hungary mathematical competition. The top scientific genius of Hungary, *J. von Neumann*, was one of the leading mathematicians in the 20th century. *Neumann* was overseas while the competition took place. Later he did it himself and it took him half an hour to complete. Another mathematician worth mentioning is the highly productive number theorist *P. Erdös*. He was a pupil of *Fejér* and also a winner of the Wolf Prize. *Erdös* was very passionate about mathematical competition and setting competition questions. His contribution to discrete mathematics was unique and greatly significant. The rapid progress and development of discrete mathematics over the subsequent decades had indirectly influenced the types of questions set in IMO. An internationally recognized prize named after Erdös was to honour those who had contributed to the education of mathematical competition. Professor *Qiu Zonghu* from China had won the prize in 1993.

In 1934, a famous mathematician *B. Delone* conducted a mathematical competition for high school students in Leningrad (now St. Petersburg). In 1935, Moscow also started organizing such event. Other than being interrupted during the World War

Ⅱ, these events had been carried on until today. As for the Russian Mathematical Competition (later renamed as the Soviet Mathematical Competition), it was not started until 1961. Thus, the former Soviet Union and Russia became the leading powers of Mathematical Olympiad. A lot of grandmasters in mathematics including the great *A. N. Kolmogorov* were all very enthusiastic about the mathematical competition. They would personally involve in setting the questions for the competition. The former Soviet Union even called it the Mathematical Olympiad, believing that mathematics is the "gymnastics of thinking". These points of view gave a great impact on the educational community. The winner of the Fields Medal in 1998, *M. Kontsevich*, was once the first runner-up of the Russian Mathematical Competition. *G. Kasparov*, the international chess grandmaster, was once the second runner-up. *Grigori Perelman*, the winner of the Fields Medal in 2006 (but he declined), who solved the Poincaré's Conjecture, was a gold medalist of IMO in 1982.

In the United States of America, due to the active promotion by the renowned mathematician *G.D. Birkhoff* and his son, together with *G. Pólya*, the Putnam mathematics competition was organized in 1938 for junior undergraduates. Many of the questions were within the scope of high school students. The top five contestants of the Putnam mathematical competition would be entitled to the membership of Putnam. Many of these were eventually outstanding mathematicians. There were the famous *R. Feynman* (winner of the Nobel Prize for Physics, 1965), *K. Wilson* (winner of the Nobel Prize for Physics, 1982), *J. Milnor* (winner of the Fields Medal, 1962), *D. Mumford* (winner of the Fields Medal, 1974), and *D.*

Quillen (winner of the Fields Medal, 1978).

Since 1972, in order to prepare for the IMO, the United States of America Mathematical Olympiad (USAMO) was organized. The standard of questions posed was very high, parallel to that of the Winter Camp in China. Prior to this, the United States had organized American High School Mathematics Examination (AHSME) for the high school students since 1950. This was at the junior level yet the most popular mathematics competition in America. Originally, it was planned to select about 100 contestants from AHSME to participate in USAMO. However, due to the discrepancy in the level of difficulty between the two competitions and other restrictions, from 1983 onwards, an intermediate level of competition, namely, American Invitational Mathematics Examination (AIME), was introduced. Henceforth both AHSME and AIME became internationally well-known. A few cities in China had participated in the competition and the results were encouraging.

The members of the national team who were selected from USAMO would undergo training at the West Point Military Academy, and would meet the President at the White House together with their parents. Similarly as in the former Soviet Union, the Mathematical Olympiad education was widely recognized in America. The book "How to Solve it" written by *George Polya* along with many other titles had been translated into many different languages. *George Polya* provided a whole series of general heuristics for solving problems of all kinds. His influence in the educational community in China should not be underestimated.

International Mathematical Olympiad

In 1956, the East European countries and the Soviet Union took the initiative to organize the IMO formally. The first International Mathematical Olympiad (IMO) was held in Brasov, Romania, in 1959. At the time, there were only seven participating countries, namely, Romania, Bulgaria, Poland, Hungary, Czechoslovakia, East Germany and the Soviet Union. Subsequently, the United States of America, United Kingdom, France, Germany and also other countries including those from Asia joined. Today, the IMO had managed to reach almost all the developed and developing countries. Except in the year 1980 due to financial difficulties faced by the host country, Mongolia, there were already 49 Olympiads held and 97 countries participating.

The mathematical topics in the IMO include number theory, polynomials, functional equations, inequalities, graph theory, complex numbers, combinatorics, geometry and game theory. These areas had provided guidance for setting questions for the competitions. Other than the first few Olympiads, each IMO is normally held in mid-July every year and the test paper consists of 6 questions in all. The actual competition lasts for 2 days for a total of 9 hours where participants are required to complete 3 questions each day. Each question is 7 points which total up to 42 points. The full score for a team is 252 marks. About half of the participants will be awarded a medal, where 1/12 will be awarded a gold medal. The numbers of gold, silver and bronze medals awarded are in the ratio of 1 : 2 : 3 approximately. In the case when a participant provides a better solution than the official answer, a special award is given.

Each participating country will take turn to host the IMO.

The cost is borne by the host country. China had successfully hosted the 31st IMO in Beijing. The event had made a great impact on the mathematical community in China. According to the rules and regulations of the IMO, all participating countries are required to send a delegation consisting of a leader, a deputy leader and 6 contestants. The problems are contributed by the participating countries and are later selected carefully by the host country for submission to the international jury set up by the host country. Eventually, only 6 problems will be accepted for use in the competition. The host country does not provide any question. The short-listed problems are subsequently translated, if necessary, in English, French, German, Russian and other working languages. After that, the team leaders will translate the problems into their own languages.

The answer scripts of each participating team will be marked by the team leader and the deputy leader. The team leader will later present the scripts of their contestants to the coordinators for assessment. If there is any dispute, the matter will be settled by the jury. The jury is formed by the various team leaders and an appointed chairman by the host country. The jury is responsible for deciding the final 6 problems for the competition. Their duties also include finalizing the grading standard, ensuring the accuracy of the translation of the problems, standardizing replies to written queries raised by participants during the competition, synchronizing differences in grading between the team leaders and the coordinators and also deciding on the cut-off points for the medals depending on the contestants' results as the difficulties of problems each year are different.

China had participated informally in the 26th IMO in 1985. Only two students were sent. Starting from 1986, except in 1998 when the IMO was held in Taiwan, China had always sent 6 official contestants to the IMO. Today, the Chinese contestants not only performed outstandingly in the IMO, but also in the International Physics, Chemistry, Informatics, and Biology Olympiads. So far, no other countries have overtaken China in the number of gold and silver medals received. This can be regarded as an indication that China pays great attention to the training of basic skills in mathematics and science education.

Winners of the IMO

Among all the IMO medalists, there were many of them who eventually became great mathematicians. They were also awarded the Fields Medal, Wolf Prize and Nevanlinna Prize (a prominent mathematics prize for computing and informatics). In what follows, we name some of the winners.

G. Margulis, a silver medalist of IMO in 1959, was awarded the Fields Medal in 1978. *L. Lovasz*, who won the Wolf Prize in 1999, was awarded the Special Award in IMO consecutively in 1965 and 1966. *V. Drinfeld*, a gold medalist of IMO in 1969, was awarded the Fields Medal in 1990. *J.-C. Yoccoz* and *T. Gowers*, who were both awarded the Fields Medal in 1998, were gold medalists in IMO in 1974 and 1981 respectively. A silver medalist of IMO in 1985, *L. Lafforgue*, won the Fields Medal in 2002. A gold medalist of IMO in 1982, *Grigori Perelman* from Russia, was awarded the Fields Medal in 2006 for solving the final step of the Poincaré conjecture. In 1986, 1987, and 1988, *Terence Tao* won a bronze, silver, and gold medal respectively. He was the youngest participant to date in the IMO, first competing at the age of ten. He was also awarded the Fields Medal in 2006. Gold medalist of IMO 1988 and 1989, *Ngo Bau Chao*, won the Fields Medal in 2010, together with the bronze medalist of IMO 1988, *E. Lindenstrauss*.

A silver medalist of IMO in 1977, *P. Shor*, was awarded the Nevanlinna Prize. A gold medalist of IMO in 1979, *A. Razborov*, was awarded the Nevanlinna Prize. Another gold medalist of IMO in 1986, *S. Smirnov*, was awarded the Clay Research Award. *V. Lafforgue*, a gold medalist of IMO in 1990, was awarded the European Mathematical Society prize. He is *L. Lafforgue's* younger brother.

Also, a famous mathematician in number theory, *N. Elkies*, who is also a professor at Harvard University, was awarded a gold medal of IMO in 1982. Other winners include *P. Kronheimer* awarded a silver medal in 1981 and *R. Taylor* a contestant of IMO in 1980.

Mathematical competition in China

Due to various reasons, mathematical competition in China started relatively late but is progressing vigorously.

"We are going to have our own mathematical competition too!" said *Hua Luogeng*. *Hua* is a house-hold name in China. The first mathematical competition was held concurrently in Beijing, Tianjin, Shanghai and Wuhan in 1956. Due to the political situation at the time, this event was interrupted a few times. Until 1962, when the political environment started to improve, Beijing and other cities started organizing the competition though not regularly. In the era of Cultural Revolution, the whole educational system in China was in chaos. The mathematical competition came to a complete halt. In contrast, the mathematical competition in the former Soviet Union was still on-going during the war and at a time under the difficult political situation. The competitions in Moscow were interrupted only 3 times between 1942 and 1944. It was indeed commendable.

In 1978, it was the spring of science. *Hua Luogeng* conducted the Middle School Mathematical Competition for 8 provinces in China. The mathematical competition in China was then making a fresh start and embarked on a road of rapid development. *Hua* passed away in 1985. In commemorating him, a competition named *Hua Luogeng* Gold Cup was set up in 1986 for students in Grade 6 and 7 and it has a great impact.

The mathematical competitions in China before 1980 can be considered as the initial period. The problems set were within the scope of middle school textbooks. After 1980, the competitions were gradually moving towards the senior middle school level. In 1981, the Chinese Mathematical Society decided to conduct the China Mathematical Competition, a national event for high schools.

In 1981, the United States of America, the host country of IMO, issued an invitation to China to participate in the event. Only in 1985, China sent two contestants to participate informally in the IMO. The results were not encouraging. In view of this, another activity called the Winter Camp was conducted after the China Mathematical Competition. The Winter Camp was later renamed as the China Mathematical Olympiad or CMO. The winning team would be awarded the *Chern Shiing-Shen* Cup. Based on the outcome at the Winter Camp, a selection would be made to form the 6-member national team for IMO. From 1986 onwards, other than the year when IMO was organized in Taiwan, China had been sending a 6-member team to IMO. Up to 2011, China had been awarded the overall team champion for 17 times.

In 1990, China had successfully hosted the 31st IMO. It showed that the standard of mathematical competition in China has leveled that of other leading countries. First, the fact that China achieves the highest marks at the 31st IMO for the team is an evidence of the effectiveness of the pyramid approach in selecting the contestants in China. Secondly, the Chinese mathematicians had simplified and modified over 100 problems and submitted them to the team leaders of the 35 countries for their perusal. Eventually, 28 problems were recommended. At the end, 5 problems were chosen (IMO requires 6 problems). This is another evidence to show that China has achieved the highest quality in setting problems. Thirdly, the answer scripts of the participants were marked by the various team leaders and assessed by the coordinators who were nominated by the host countries. China had formed a group 50 mathematicians to serve as coordinators who would ensure the high accuracy and fairness in marking. The marking process was completed half a day earlier than it was scheduled. Fourthly, that was the first ever IMO organized in Asia. The outstanding performance by China had encouraged the other developing countries, especially those in Asia. The organizing and coordinating work of the IMO by the host country was also reasonably good.

In China, the outstanding performance in mathematical competition is a result of many contributions from the all quarters of mathematical community. There are the older generation of mathematicians, middle-aged mathematicians and also the middle and elementary school teachers. There is one person who deserves a special mention and he is *Hua Luogeng*. He initiated and promoted the mathematical competition. He is also the author of the following books: Beyond *Yang hui's* Triangle, Beyond the *pi* of *Zu Chongzhi*, Beyond the Magic Computation of *Sun-zi*, Mathematical Induction, and Mathematical Problems of Bee Hive. These were his books derived from mathematics competitions. When China resumed mathematical competition in 1978, he participated in setting problems and giving critique to solutions of the problems. Other outstanding books derived from the Chinese mathematics competitions are: Symmetry by *Duan Xuefu*, Lattice and Area by *Min Sihe*, One Stroke Drawing and Postman Problem by *Jiang Boju*.

After 1980, the younger mathematicians in China had taken over from the older generation of mathematicians in running the mathematical competition. They worked and strived hard to bring the level of mathematical competition in China to a new height. *Qiu Zonghu* is one such outstanding representative. From the training of contestants and leading the team 3 times to IMO to the organizing of the 31th IMO in China, he had contributed prominently and was awarded the *P. Erdös* prize.

Preparation for IMO

Currently, the selection process of participants for IMO in China is as follows.

First, the China Mathematical Competition, a national competition for high Schools, is organized on the second Sunday in October every year. The objectives are: to increase the interest of students in learning mathematics, to promote the development of co-curricular activities in mathematics, to help improve the teaching of mathematics in high schools, to discover and cultivate the talents and also to prepare for the IMO. This happens since 1981. Currently there are about 200,000 participants taking part.

Through the China Mathematical Competition, around 150 of students are selected to take part in the China Mathematical Olympiad or CMO, that is, the Winter Camp. The CMO lasts for 5 days and is held in January every year. The types and difficulties of the problems in CMO are very much similar to the IMO. There are also 3 problems to be completed within 4.5 hours each day. However, the score for each problem is 21 marks which add up to 126 marks in total. Starting from 1990, the Winter Camp instituted the *Chern Shiing-Shen* Cup for team championship. In 1991, the Winter Camp was officially renamed as the China Mathematical Olympiad (CMO). It is similar to the highest national mathematical competition in the former Soviet Union and the United States.

The CMO awards the first, second and third prizes. Among the participants of CMO, about 20 to 30 students are selected to participate in the training for IMO. The training takes place in March every year. After 6 to 8 tests and another 2 rounds of qualifying examinations, only 6 contestants are short-listed to form the China IMO national team to take part in the IMO in July.

Besides the China Mathematical Competition (for high schools), the Junior Middle School Mathematical Competition is also developing well. Starting from 1984, the competition is organized in April every year by the Popularization Committee of the Chinese Mathematical Society. The various provinces, cities and autonomous regions would rotate to host the event. Another mathematical competition for the junior middle schools is also conducted in April every year by the Middle School Mathematics Education Society of the Chinese Educational Society since 1998 till now.

The *Hua Luogeng* Gold Cup, a competition by invitation, had also been successfully conducted since 1986. The participating students comprise elementary six and junior middle one students. The format of the competition consists of a preliminary round, semifinals in various provinces, cities and autonomous regions, then the finals.

Mathematical competition in China provides a platform for students to showcase their talents in mathematics. It encourages learning of mathematics among students. It helps identify talented students and to provide them with differentiated learning opportunity. It develops co-curricular activities in mathematics. Finally, it brings about changes in the teaching of mathematics.

Contents

China Mathematical Competition

2008 (Chongqing)

The Popularization Committee of Chinese Methematica Society and Chongqing Mathematical Society were responsible for the assignment of the competition problems in the first round test and the supplementary test.

Part I Multiple-choice Questions (Questions 1—6, six marks each)

1 The minimum value of $f(x)=\dfrac{5-4x+x^2}{2-x}$ for $x \in (-\infty, 2)$ is ( ).

(A) 0 (B) 1 (C) 2 (D) 3

Solution Let $x < 2 \Rightarrow 2 - x > 0$. Then

$$f(x) = \frac{1 + (4 - 4x + x^2)}{2 - x} = \frac{1}{2 - x} + (2 - x)$$
$$\geqslant 2 \times \sqrt{\frac{1}{2 - x} \times (2 - x)} = 2.$$

The equality holds if and only if $\frac{1}{2 - x} = 2 - x$, and it is so when $x = 1 \in (-\infty, 2)$. This means that f reaches the minimum value 2 at $x = 1$.

Answer: C

2 Let $A = [-2, 4)$, $B = \{x \mid x^2 - ax - 4 \leqslant 0\}$. If $B \subseteq A$, then the range of real a is ( ).

(A) $[-1, 2)$ (B) $[-1, 2]$

(C) $[0, 3]$ (D) $[0, 3)$

Solution $x^2 - ax - 4 = 0$ has two roots:

$$x_1 = \frac{a}{2} - \sqrt{4 + \frac{a^2}{4}}, \quad x_2 = \frac{a}{2} + \sqrt{4 + \frac{a^2}{4}}.$$

We have $B \subseteq A \Leftrightarrow x_1 \geqslant -2$ and $x_2 < 4$. This means that

$$\frac{a}{2} - \sqrt{4 + \frac{a^2}{4}} \geqslant -2, \quad \frac{a}{2} + \sqrt{4 + \frac{a^2}{4}} < 4.$$

From the above we get $0 \leqslant a < 3$.

Answer: D

3 A and B are playing ping-pong, with the agreement that the winner of a game will get 1 point and the loser 0 point; the match ends as soon as one of the players is ahead by 2 points or the number of games reaches six. Suppose that the probabilities of A and B winning a game are $\frac{2}{3}$ and $\frac{1}{3}$, respectively, and each game is independent. Then the expectation $E\xi$ for the match ending with ξ games is ( ).

(A) $\frac{241}{81}$ (B) $\frac{266}{81}$ (C) $\frac{274}{81}$ (D) $\frac{670}{243}$

Solution I It is easy to see that ξ can only be 2, 4 or 6. We divide the six games into three rounds, each consisting of two consecutive games. If one of the players wins two games in the first round, the match ends and the probability is

$$\left(\frac{2}{3}\right)^2+\left(\frac{1}{3}\right)^2=\frac{5}{9}.$$

Otherwise the players tie with each other, earning one point each, and the match enters the second round; this probability is $1-\frac{5}{9}=\frac{4}{9}$.

We have similar discussions for the second and third rounds. So we get

$$P(\xi=2)=\frac{5}{9},$$

$$P(\xi=4)=\frac{4}{9}\times\frac{5}{9}=\frac{20}{81},$$

$$P(\xi=6)=\left(\frac{4}{9}\right)^2=\frac{16}{81}.$$

Then

$$E\xi=2\times\frac{5}{9}+4\times\frac{20}{81}+6\times\frac{16}{81}=\frac{266}{81}.$$

Answer: B

Solution II Let A_k denote the event that A wins the kth game, while $\overline{A}_k$ means that B wins the game. Since A_k and $\overline{A}_k$ are incompatible, and are independent of the other events, we have

$$P(\xi = 2) = P(A_1A_2) + P(\overline{A}_1\overline{A}_2) = \frac{5}{9},$$

$$\begin{aligned} P(\xi = 4) &= P(A_1\overline{A}_2A_3A_4) + P(A_1\overline{A}_2\overline{A}_3\overline{A}_4) + \\ &\quad P(\overline{A}_1A_2A_3A_4) + P(\overline{A}_1A_2\overline{A}_3\overline{A}_4) \\ &= 2\left[\left(\frac{2}{3}\right)^3\left(\frac{1}{3}\right) + \left(\frac{1}{3}\right)^3\left(\frac{2}{3}\right)\right] = \frac{20}{81}, \end{aligned}$$

$$\begin{aligned} P(\xi = 6) &= P(A_1\overline{A}_2A_3\overline{A}_4) + P(A_1\overline{A}_2\overline{A}_3A_4) + \\ &\quad P(\overline{A}_1A_2A_3\overline{A}_4) + P(\overline{A}_1A_2\overline{A}_3A_4) \\ &= 4\left(\frac{2}{3}\right)^2\left(\frac{1}{3}\right)^2 = \frac{16}{81}. \end{aligned}$$

Then

$$E\xi = 2 \times \frac{5}{9} + 4 \times \frac{20}{81} + 6 \times \frac{16}{81} = \frac{266}{81}.$$

Answer: B

4 Given three cubes with integer edge lengths, if the sum of their surface areas is 564 cm^2, then the sum of their volumes is ( ).

(A) 764 cm^3 or 586 cm^3 (B) 764 cm^3

(C) 586 cm^3 or 564 cm^3 (D) 586 cm^3

Solution Denote the edge lengths of the three cubes as a, b and c, respectively. Then we have

$$6(a^2 + b^2 + c^2) = 564,$$

i.e. $a^2 + b^2 + c^2 = 94$. We may assume that

$$1 \leqslant a \leqslant b \leqslant c < 10.$$

Then

$$3c^2 \geqslant a^2 + b^2 + c^2 = 94.$$

It follows that $c^2 > 31$. So $6 \leqslant c < 10$, and this means that c can only be 9, 8, 7 or 6.

If $c = 9$, then

$$a^2 + b^2 = 94 - 9^2 = 13.$$

It is easy to see that $a = 2$, $b = 3$. So we get the solution $(a, b, c) = (2, 3, 9)$.

If $c = 8$, then

$$a^2 + b^2 = 94 - 8^2 = 30.$$

This means that $b \geqslant 4$ and $2b^2 \geqslant 30$; it follows that $b = 4$ or 5, so $a^2 = 5$ or 14; in both cases a has no integer solution.

If $c = 7$, then

$$a^2 + b^2 = 94 - 7^2 = 45.$$

It is easy to see that $a = 3$, $b = 6$ is the only solution.

If $c = 6$, then

$$a^2 + b^2 = 94 - 6^2 = 58.$$

So $2b^2 \geqslant 58$, or $b^2 \geqslant 29$. This means that $b \geqslant 6$, but $b \leqslant c = 6$, so $b = 6$. Then $a^2 = 22$ and a cannot be an integer.

In summary, there are two solutions: $(a, b, c) = (2, 3, 9)$ and $(a, b, c) = (3, 6, 7)$. Then the possible volumes are

$$V_1 = 2^3 + 3^3 + 9^3 = 764 \text{ cm}^3,$$
$$V_2 = 3^3 + 6^3 + 7^3 = 586 \text{ cm}^3.$$

Answer: A

5 The number of rational solutions to the system of equations

$$\begin{cases} x+y+z=0, \\ xyz+z=0, \\ xy+yz+xz+y=0 \end{cases}$$ is ( ).

(A) 1 (B) 2 (C) 3 (D) 4

Solution If $z=0$, then $\begin{cases} x+y=0, \\ xy+y=0. \end{cases}$ It follows that

$$\begin{cases} x=0, \\ y=0 \end{cases} \text{or} \begin{cases} x=-1, \\ y=1. \end{cases}$$

If $z \neq 0$, from $xyz+z=0$ we get

$$xy=-1. \quad ①$$

From $x+y+z=0$ we have

$$z=-x-y. \quad ②$$

Substituting ② into $xy+yz+xz+y=0$, we obtain

$$x^2+y^2+xy-y=0. \quad ③$$

From ① we have $x=-\dfrac{1}{y}$. We substitute it into ③, and then make a simplification:

$$(y-1)(y^3-y-1)=0.$$

It is easy to see that y^3-y-1 has no rational solution, and so $y=1$. Then from ① and ② we get $x=-1$ and $z=0$, contradicting $z \neq 0$.

In summary, the system has exactly two solutions:

$$\begin{cases} x=0, \\ y=0, \\ z=0, \end{cases} \begin{cases} x=-1, \\ y=1, \\ z=0. \end{cases}$$

Answer: B

6 Suppose that the sides a, b, c of $\triangle ABC$, corresponding to the angles A, B, C respectively, constitute a geometric sequence. Then the range of $\frac{\sin A\cot C+\cos A}{\sin B\cot C+\cos B}$ is ( ).

(A) $(0, +\infty)$ (B) $\left(0, \frac{\sqrt{5}+1}{2}\right)$

(C) $\left(\frac{\sqrt{5}-1}{2}, \frac{\sqrt{5}+1}{2}\right)$ (D) $\left(\frac{\sqrt{5}-1}{2}, +\infty\right)$

Solution Suppose that the common ratio of a, b, c is q. Then $b = aq$, $c = aq^2$. We have

$$\begin{aligned}\frac{\sin A\cot C+\cos A}{\sin B\cot C+\cos B} &= \frac{\sin A\cos C+\cos A\sin C}{\sin B\cos C+\cos B\sin C} \\ &= \frac{\sin(A+C)}{\sin(B+C)} = \frac{\sin(\pi-B)}{\sin(\pi-A)} \\ &= \frac{\sin B}{\sin A} = \frac{b}{a} = q.\end{aligned}$$

So we only need to determine the range of q. As a, b, c are the sides of a triangle, they satisfy $a+b>c$ and $b+c>a$. That is to say,

$$\begin{cases} a+aq>aq^2, \\ aq+aq^2>a. \end{cases}$$

It follows that

$$\begin{cases} q^2-q-1<0, \\ q^2+q-1>0. \end{cases}$$

Their solutions are

$$\begin{cases} \frac{\sqrt{5}-1}{2}<q<\frac{\sqrt{5}+1}{2}, \\ q>\frac{\sqrt{5}-1}{2} \text{ or } q<-\frac{\sqrt{5}+1}{2}. \end{cases}$$

It is only possible that

$$\frac{\sqrt{5}-1}{2} < q < \frac{\sqrt{5}+1}{2}.$$

Answer: C

Part II Short-Answer Questions (Questions 7—12, nine marks each)

7 Let $f(x) = ax + b$, with a, b real numbers; $f_1(x) = f(x)$, $f_{n+1}(x) = f(f_n(x))$, $n = 1, 2, \cdots$. If $f_7(x) = 128x + 381$, then $a + b =$ ________.

Solution From the definitions we get

$$f_n(x) = a^n x + (a^{n-1} + a^{n-2} + \cdots + a + 1)b$$
$$= a^n x + \frac{a^n - 1}{a - 1} \times b.$$

As $f_7(x) = 128x + 381$, we have $a^7 = 128$ and $\frac{a^7 - 1}{a - 1} \times b = 381$. Then $a = 2$, $b = 3$. The answer is $a + b = 5$.

8 Suppose that the minimum of $f(x) = \cos 2x - 2a(1 + \cos x)$ is $-\frac{1}{2}$. Then $a =$ ________.

Solution We have

$$f(x) = 2\cos x^2 - 1 - 2a - 2a\cos x$$
$$= 2\left(\cos x - \frac{a}{2}\right)^2 - \frac{1}{2}a^2 - 2a - 1.$$

For $a > 2$, $f(x)$ takes the minimum value of $1 - 4a$ when $\cos x = 1$; for $a < -2$, $f(x)$ takes the minimum 1 when $\cos x = -1$; for $-2 \leqslant a \leqslant 2$, $f(x)$ takes the minimum $-\frac{1}{2}a^2 - 2a - 1$

when $\cos x = \frac{a}{2}$. It is easy to see that $f(x)$ will never be $-\frac{1}{2}$ for $a > 2$ or $a < -2$. So it is only possible that $-2 \leqslant a \leqslant 2$. Then from $-\frac{1}{2}a^2 - 2a - 1 = \frac{1}{2}$, we get $a = -2 + \sqrt{3}$ or $a = -2 - \sqrt{3}$ (discarded). Therefore, the correct answer is $a = -2 + \sqrt{3}$.

9 Twenty-four volunteers will be allocated to three schools. The rule is that each school will accept at least one volunteer and all the schools will accept different numbers of volunteers. Then there are ________ different ways of allocating volunteers.

Solution We may use each space between every two consecutive bars (|) to represent a school and each asterisk (*) to represent a volunteer, as seen in the following example: the first, second and third schools receive 4, 18 and 2 volunteers, respectively.

| * * * * | * ··· * | * * |

Then the allocation problem may be regarded as a permutation-and-combination problem of 4 bars and 24 asterisks.

Since the two ends of the line must be occupied by a bar, respectively, there are $\binom{23}{2} = 253$ ways to insert the other 2 bars into the 23 spaces between the 24 asterisks such that there is at least 1 asterisk between every two consecutive bars, in which there are 31 ways that at least two schools have the same number of volunteers. So the number of allocating ways satisfying the conditions is $253 - 31 = 222$.

10 Let S_n denote the sum of the first n terms in a number sequence $\{a_n\}$, satisfying

$$S_n + a_n = \frac{n-1}{n(n+1)},\ n = 1, 2, \cdots.$$

Then $a_n =$ ________.

Solution As

$$\begin{aligned} a_{n+1} &= S_{n+1} - S_n \\ &= \frac{n}{(n+1)(n+2)} - a_{n+1} - \frac{n-1}{n(n+1)} + a_n, \end{aligned}$$

we have

$$\begin{aligned} 2a_{n+1} &= \frac{n+2-2}{(n+1)(n+2)} - \frac{1}{n+1} + \frac{1}{n(n+1)} + a_n \\ &= \frac{-2}{(n+1)(n+2)} + a_n + \frac{1}{n(n+1)}. \end{aligned}$$

Therefore,

$$a_{n+1} + \frac{1}{(n+1)(n+2)} = \frac{1}{2}\left(a_n + \frac{1}{n(n+1)}\right).$$

Define $b_n = a_n + \frac{1}{n(n+1)}$. It is easy to see that $b_n = \frac{1}{2^{n-1}}b_1$, $b_1 = a_1 + \frac{1}{2}$. On the other hand, from $S_1 + a_1 = 2a_1 = 0$ we get $a_1 = 0$. So $b_1 = \frac{1}{2}$, $b_n = \frac{1}{2^n}$. Therefore,

$$a_n = b_n - \frac{1}{n(n+1)} = \frac{1}{2^n} - \frac{1}{n(n+1)}.$$

11 Suppose that $f(x)$ is defined on $\mathbf{R}$, satisfying $f(0) = 2008$, and for any $x \in \mathbf{R}$

$$f(x+2) - f(x) \leqslant 3 \times 2^x,$$
$$f(x+6) - f(x) \geqslant 63 \times 2^x.$$

Then $f(2008) =$ ________.

Solution I We have

$$\begin{aligned}
& f(x+2)-f(x) \\
=& -(f(x+4)-f(x+2))-(f(x+6) \\
& -f(x+4))+(f(x+6)-f(x)) \\
\geqslant & -3\times 2^{x+2}-3\times 2^{x+4}+63\times 2^{x}=3\times 2^{x}.
\end{aligned}$$

This means that $f(x+2)-f(x)=3\times 2^{x}$. So we have

$$\begin{aligned}
f(2008) &= f(2008)-f(2006)+f(2006)-f(2004)+\cdots \\
&\quad +f(2)-f(0)+f(0) \\
&= 3\times(2^{2006}+2^{2004}+\cdots+2^{2}+1)+f(0) \\
&= 3\times\frac{4^{1003+1}-1}{4-1}+2008 \\
&= 2^{2008}+2007.
\end{aligned}$$

Solution II We define $g(x)=f(x)-2^{x}$. Then we have

$$\begin{aligned}
g(x+2)-g(x) &= f(x+2)-f(x)-2^{x+2}+2^{x} \\
&\leqslant 3\times 2^{x}-3\times 2^{x}=0, \\
g(x+6)-g(x) &= f(x+6)-f(x)-2^{x+6}+2^{x} \\
&\geqslant 63\times 2^{x}-63\times 2^{x}=0.
\end{aligned}$$

This means that $g(x)\leqslant g(x+6)\leqslant g(x+4)\leqslant g(x+2)\leqslant g(x)$, and it implies that $g(x)$ is a periodic function with 2 as a period. So

$$\begin{aligned}
f(2008) &= g(2008)+2^{2008}=g(0)+2^{2008} \\
&= 2007+2^{2008}.
\end{aligned}$$

12 Suppose that a ball with radius 1 moves freely inside a regular tetrahedron with edge length $4\sqrt{6}$. Then the area of the inner surface of the container, which the ball can never touch, is ________.

Solution As shown in Fig. 1, consider the situation where the ball is in a corner of the container. Draw the plane $A_1B_1C_1 \parallel ABC$, tangent to the ball at point D. Then the ball center O is also the center of the tetrahedron $P\text{-}A_1B_1C_1$, with $PO \perp A_1B_1C_1$ and the foot point D being the center of $\triangle A_1B_1C_1$.

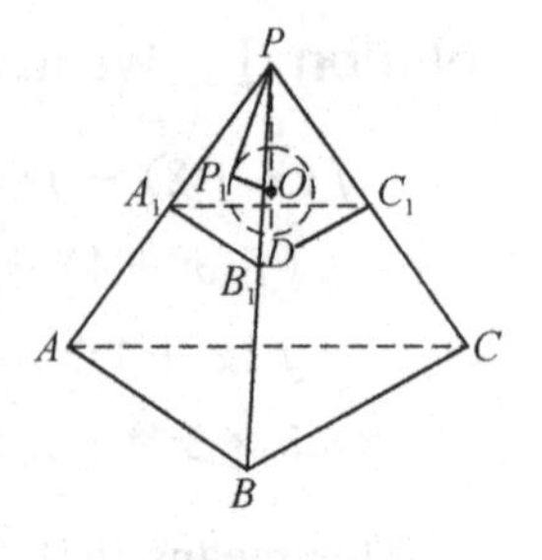

Fig. 1

Since

$$\begin{aligned} V_{P-A_1B_1C_1} &= \frac{1}{3} \times S_{\triangle A_1B_1C_1} \times PD \\ &= 4 \times V_{O-A_1B_1C_1} \\ &= 4 \times \frac{1}{3} \times S_{\triangle A_1B_1C_1} \times OD, \end{aligned}$$

we have $PD = 4OD = 4r$, where r is the radius of the ball. It follows that

$$PO = PD - OD = 3r.$$

Suppose that the ball is tangent to the plane PAB at point P_1. Then we have

$$PP_1 = \sqrt{PO^2 - OP_1^2} = \sqrt{(3r)^2 - r^2} = 2\sqrt{2}r.$$

As shown in Fig. 2, it is easy to see that the locus of the ball on the plane PAB is also a regular triangle, denoted by P_1EF. Through P_1 draw $P_1M \perp PA$ with point M on PA. Then $\angle MPP_1 = \frac{\pi}{6}$, and

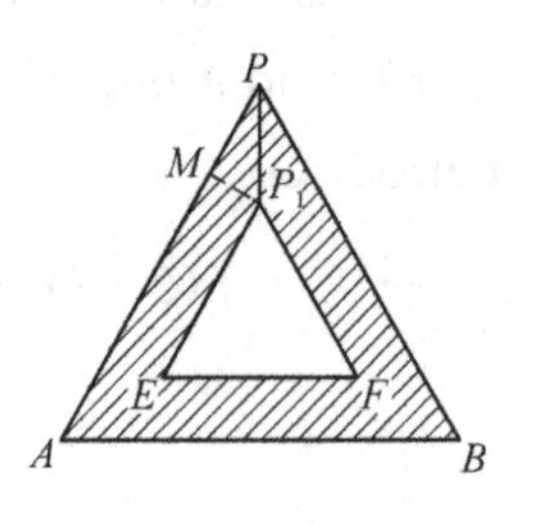

Fig. 2

$$PM = PP_1 \times \cos\angle MPP_1 = 2\sqrt{2}r \times \frac{\sqrt{3}}{2} = \sqrt{6}r.$$

It follows that $P_1E = PA - 2PM = a - 2\sqrt{6}r$, where $a = PA$. Now, the space on PAB which the ball will never touch is the shaded part of Fig. 2, and its size is equal to

$$
\begin{aligned}
S_{\triangle PAB} - S_{\triangle P_1EF} &= \frac{\sqrt{3}}{4}(a^2 - (a - 2\sqrt{6}r)^2) \\
&= 3\sqrt{2}ar - 6\sqrt{3}r^2 \\
&= 24\sqrt{3} - 6\sqrt{3} \\
&= 18\sqrt{3},
\end{aligned}
$$

since $r = 1$ and $a = 4\sqrt{6}$ under given conditions. Then the total untouched area is

$$4 \times 18\sqrt{3} = 72\sqrt{3}.$$

Part III Word Problems (Questions 13—15, 20 marks each)

13 It is known that the curve $f(x) = |\sin x|$ intercepts the line $y = kx$ $(k > 0)$ at exactly three points, the maximum x coordinate of these points being α. Prove that

$$\frac{\cos\alpha}{\sin\alpha + \sin 3\alpha} = \frac{1 + \alpha^2}{4\alpha}.$$

Solution The image of the three intercepting points of $f(x)$ and $y = kx$ is shown in the figure. It is easy to see that the curve and the line are tangent to each other at point $A(\alpha, -\sin\alpha)$, and $\alpha \in \left(\pi, \frac{3\pi}{2}\right)$.

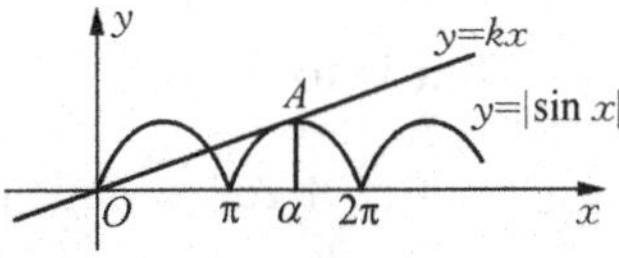

As $f'(x) = -\cos x$ for $x \in \left(\pi, \frac{3\pi}{2}\right)$, we have $-\cos\alpha = -\frac{\sin\alpha}{\alpha}$, i. e. $\alpha = \tan\alpha$. Then

$$
\begin{aligned}
\frac{\cos\alpha}{\sin\alpha + \sin 3\alpha} &= \frac{\cos\alpha}{2\sin 2\alpha\cos\alpha} = \frac{1}{4\sin\alpha\cos\alpha} \\
&= \frac{\cos^2\alpha + \sin^2\alpha}{4\sin\alpha\cos\alpha} = \frac{1 + \tan^2\alpha}{4\tan\alpha} \\
&= \frac{1 + \alpha^2}{4\alpha}.
\end{aligned}
$$

14 Solve the inequality

$$\log_2(x^{12}+3x^{10}+5x^8+3x^6+1)<1+\log_2(x^4+1).$$

Solution I As

$$1+\log_2(x^4+1)=\log_2(2x^4+2),$$

and $\log_2 y$ is monotonically increasing over $(0,+\infty)$, the given inequality is equivalent to

$$x^{12}+3x^{10}+5x^8+3x^6+1<2x^4+2$$

or

$$x^{12}+3x^{10}+5x^8+3x^6-2x^4-1<0.$$

It can be rewritten as

$$\begin{aligned}&x^{12}+x^{10}-x^8\\&\quad+2x^{10}+2x^8-2x^6\\&\qquad+4x^8+4x^6-4x^4\\&\qquad\quad+x^6+x^4-x^2\\&\qquad\qquad+x^4+x^2-1<0.\end{aligned}$$

That is to say,

$$(x^8+2x^6+4x^4+x^2+1)(x^4+x^2-1)<0.$$

Then we have $x^4+x^2-1<0$. It follows that $x^2<\frac{-1+\sqrt{5}}{2}$, i.e.

$$-\sqrt{\frac{-1+\sqrt{5}}{2}}<x<\sqrt{\frac{-1+\sqrt{5}}{2}}.$$

So the solution set is $\left(-\sqrt{\frac{-1+\sqrt{5}}{2}},\sqrt{\frac{-1+\sqrt{5}}{2}}\right)$.

Solution II As

$$1+\log_2(x^4+1)=\log_2(2x^4+2),$$

and $\log_2 y$ is monotonically increasing over $(0,+\infty)$, the given inequality is equivalent to

$$x^{12}+3x^{10}+5x^8+3x^6+1<2x^4+2$$

or

$$\left(\frac{1}{x^2}\right)^3+2\left(\frac{1}{x^2}\right)>x^6+3x^4+3x^2+1+2x^2+2$$
$$=(x^2+1)^3+2(x^2+1).$$

Define $g(t)=t^2+2t$. Then we have

$$g\left(\frac{1}{x^2}\right)>g(x^2+1).$$

Obviously, $g(t)$ is a monotonically increasing function; then we have

$$\frac{1}{x^2}>x^2+1.$$

That is to say,

$$x^4+x^2-1<0.$$

We obtain $x^2<\frac{-1+\sqrt{5}}{2}$. So the solution set is $\left(-\sqrt{\frac{-1+\sqrt{5}}{2}},\sqrt{\frac{-1+\sqrt{5}}{2}}\right)$.

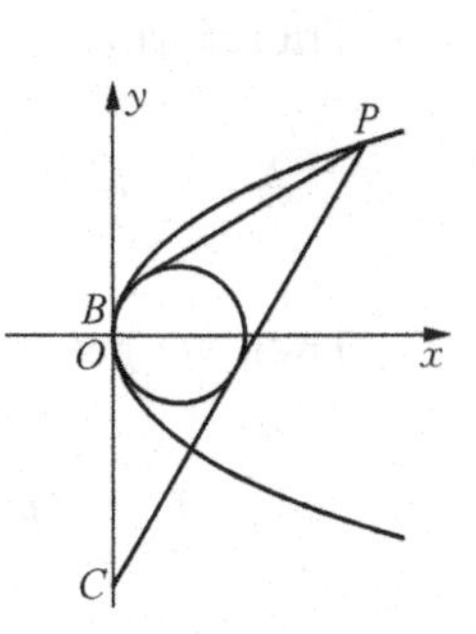

15 As is shown in the figure, P is a moving point on the parabola $y^2=2x$, points B, C are on the y axis, and the circle $(x-1)^2+y^2=1$ is internally tangent to $\triangle PBC$. Find the minimum value of the area of $\triangle PBC$.

Solution Denote P, B, C by $P(x_0, y_0)$, $B(0, b)$, $C(0, c)$, and assume that $b > c$. The equation for the line PB is

$$y - b = \frac{y_0 - b}{x_0} x.$$

It can be rewritten as

$$(y_0 - b)x - x_0 y + x_0 b = 0.$$

Since the distance between the circle center $(1, 0)$ and the line PB is 1, we have

$$\frac{| y_0 - b + x_0 b |}{\sqrt{(y_0 - b)^2 + x_0^2}} = 1.$$

That is to say,

$$(y_0 - b)^2 + x_0^2 = (y_0 - b)^2 + 2x_0 b(y_0 - b) + x_0^2 b^2.$$

It is easy to see that $x_0 > 2$. Then the last equation can be simplified as

$$(x_0 - 2)b^2 + 2y_0 b - x_0 = 0.$$

In a similar way,

$$(x_0 - 2)c^2 + 2y_0 c - x_0 = 0.$$

Therefore,

$$b + c = \frac{-2y_0}{x_0 - 2}, \quad bc = \frac{-x_0}{x_0 - 2}.$$

Then we get

$$(b - c)^2 = \frac{4x_0^2 + 4y_0^2 - 8x_0}{(x_0 - 2)^2}.$$

As $P(x_0, y_0)$ is on the parabola, $y_0^2 = 2x_0$. So we have

$$(b-c)^2 = \frac{4x_0^2}{(x_0-2)^2},$$

or $b - c = \frac{2x_0}{x_0-2}$. Then we have

$$\begin{aligned} S_{\triangle PBC} &= \frac{1}{2}(b-c) \times x_0 \\ &= \frac{x_0}{x_0-2} \times x_0 \\ &= (x_0-2) + \frac{4}{x_0-2} + 4 \\ &\geqslant 4 + 4 = 8. \end{aligned}$$

The equality holds when $x_0 - 2 = 2$; this means that $x_0 = 4$ and $y_0 = \pm 2\sqrt{2}$. So the minimum of $S_{\triangle PBC}$ is 8.

2009 (Heilongjiang)

The Popularization Committee of CMS and the Mathematical Society of Heilongjiang Province were responsible for the assignment of the competition problems in the first round test and the extra test.

Part I Short-Answer Questions (Questions 1—8, seven marks each)

1 Suppose that $f(x) = \frac{x}{\sqrt{1+x^2}}$ and $f^{(n)}(x) = \underbrace{f[f[f \ldots f}_{n}(x)]]$. Then $f^{(99)}(1) =$ ________.

Solution We have

$$f^{(1)}(x) = f(x) = \frac{x}{\sqrt{1+x^2}},$$

$$f^{(2)}(x) = f[f(x)] = \frac{x}{\sqrt{1+2x^2}},$$

$$\cdots$$

$$f^{(99)}(x) = \frac{x}{\sqrt{1+99x^2}}.$$

Therefore, $f^{(99)}(1) = \frac{1}{10}$.

2 Given the line L: $x + y - 9 = 0$ and the circle M: $2x^2 + 2y^2 - 8x - 8y - 1 = 0$, point A is on L and points B, C are on M; $\angle BAC = 45°$ and the line AB is through the center of M. Then the range of the x coordinate of point A is ________.

Solution Suppose that $A(a, 9-a)$. Then the distance from the center of M to the line AC is

$$\begin{aligned} d &= |AM| \times \sin\angle BAC \\ &= \sqrt{(a-2)^2 + (9-a-2)^2} \times \sin 45° \\ &= \sqrt{2a^2 - 18a + 53} \times \frac{\sqrt{2}}{2}. \end{aligned}$$

On the other hand, since the line AC intercepts M, it follows that $d \leqslant$ the radius of $M = \sqrt{\frac{17}{2}}$, i.e.

$$\sqrt{2a^2 - 18a + 53} \times \frac{\sqrt{2}}{2} \leqslant \sqrt{\frac{17}{2}}.$$

The solution is $3 \leqslant a \leqslant 6$.

3 On a coordinate plane there are two regions, M and N: M is confined by $\begin{cases} y \geqslant 0, \\ y \leqslant x, \\ y \leqslant 2-x; \end{cases}$ and N is determined by the inequalities $t \leqslant x \leqslant t+1$, $0 \leqslant t \leqslant 1$. Then the size of the common area of M and N is given by $f(t) =$ _______.

Solution As shown in the figure, we have

$$\begin{aligned} f(t) &= S_{\text{shaded area}} \\ &= S_{\triangle AOB} - S_{\triangle OCD} - S_{\triangle BEF} \\ &= 1 - \frac{1}{2}t^2 - \frac{1}{2}(1-t)^2 \\ &= -t^2 + t + \frac{1}{2},\ 0 \leqslant t \leqslant 1. \end{aligned}$$

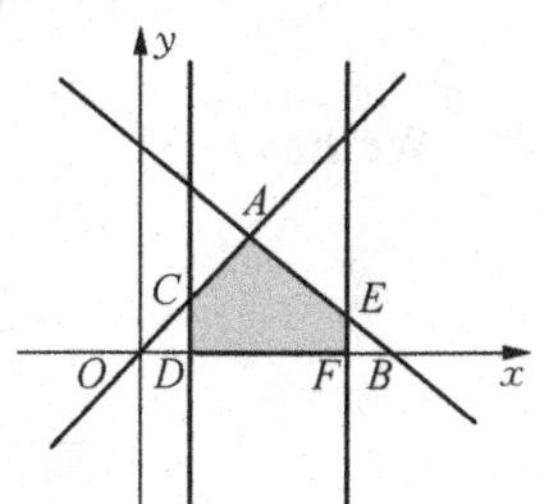

4 The inequality

$$\frac{1}{n+1} + \frac{1}{n+2} + \cdots + \frac{1}{2n+1} < a - 2007\frac{1}{3}$$

holds for every positive integer n. Then the least positive integer of a is _______.

Solution Obviously,

$$f(n) = \frac{1}{n+1} + \frac{1}{n+2} + \cdots + \frac{1}{2n+1}$$

is monotonically decreasing. Therefore, $f(1)$ reaches the maximum of $f(n)$. From

$$f(1) = \frac{1}{2} + \frac{1}{3} < a - 2007\frac{1}{3},$$

we have $a > 2008$. Therefore, the least positive integer of a is 2009.

5 Given points P, Q on an ellipse $\frac{x^2}{a^2}+\frac{y^2}{b^2}=1$ $(a>b>0)$, satisfying $OP \perp OQ$, the minimum of $|OP| \times |OQ|$ is ______.

Solution Define

$$P(|OP|\cos\theta, |OP|\sin\theta),$$
$$Q\left(|OQ|\cos\left(\theta \pm \frac{\pi}{2}\right), |OQ|\sin\left(\theta \pm \frac{\pi}{2}\right)\right).$$

We have

$$\frac{1}{|OP|^2}=\frac{\cos^2\theta}{a^2}+\frac{\sin^2\theta}{b^2}, \quad ①$$
$$\frac{1}{|OQ|^2}=\frac{\sin^2\theta}{a^2}+\frac{\cos^2\theta}{b^2}. \quad ②$$

Then

$$\frac{1}{|OP|^2}+\frac{1}{|OQ|^2}=\frac{1}{a^2}+\frac{1}{b^2}.$$

Therefore, $|OP| \times |OQ|$ reaches the minimum $\frac{2a^2b^2}{a^2+b^2}$ when $|OP|=|OQ|=\sqrt{\frac{2a^2b^2}{a^2+b^2}}$.

6 Suppose that the equation $\lg kx = 2\lg(x+1)$ has exactly one real root. Then the range of k is ______.

Solution We have

$$kx>0, \quad ①$$
$$x+1>0, \quad ②$$
$$kx=(x+1)^2. \quad ③$$

The expression ③ can be standardized as

$$x^2+(2-k)x+1=0. \quad ④$$

The two roots of ④ are

$$x_1, x_2 = \frac{1}{2}[k - 2 \pm \sqrt{k^2 - 4k}], \quad ⑤$$

where

$$\Delta = k^2 - 4k \geqslant 0 \Leftrightarrow k \leqslant 0 \text{ or } k \geqslant 4.$$

(i) When $k < 0$, it is easy to see from ⑤ that $x_1 + 1 > 0$, $x_2 + 1 < 0$, and $kx_1 > 0$. Then the equation has one real root, $x_1 = \frac{1}{2}[k - 2 + \sqrt{k^2 - 4k}]$.

(ii) When $k = 4$, the equation has one real root, $x = \frac{k}{2} - 1 = 1$.

(iii) When $k > 4$, the two roots x_1, x_2 are both positive, as well as $x_1 \neq x_2$. Discarded.

Therefore, the range of k is $k < 0$, $k = 4$.

7 Consider a pattern of numbers in the shape of a triangle: the first row consists of numbers from 1 to 100 arranged in order; each number in the second row is the sum of the two numbers directly below it in the first row; each number in the third row is the sum of the two numbers directly below it in the second row; ....

The number in the last row is ________.

Solution It is easy to see that

(i) There are 100 rows in the pattern;

(ii) The numbers in the ith row constitute an arithmetic sequence with the common difference

$$d_i = 2^{i-1}, \ i = 1, 2, \ldots, 99.$$

(iii) We have

$$\begin{aligned} a_n &= a_{n-1} + (a_{n-1} + 2^{n-2}) \\ &= 2a_{n-1} + 2^{n-2} \\ &= 2[2a_{n-2} + 2^{n-3}] + 2^{n-2} \\ &= 2^2[2a_{n-3} + 2^{n-4}] + 2 \times 2^{n-2} \\ &\cdots \\ &= 2^{n-1}a_1 + (n-1) \times 2^{n-2} \\ &= (n+1)2^{n-2}. \end{aligned}$$

Therefore, the number in the last row is $a_{100} = 101 \times 2^{98}$.

8 Every day at a railway station, there is just one train arriving between 8:00 am and 9:00 am and between 9:00 am and 10:00 am, respectively. The arrival times and their probabilities for the two trains are shown in the following table:

Arrival time	Train A	8:10	8:30	8:50
	Train B	9:10	9:30	9:50
Probability		$\frac{1}{6}$	$\frac{1}{2}$	$\frac{1}{3}$

Suppose that these random events are independent of each other. Now, a traveler comes into the station at 8:20. Then the mathematical expectation of his waiting time is ________ (round to minute).

Solution The distribution table for the waiting times of the traveler is shown below.

Waiting time/min	10	30	50	70	90
Probability	$\frac{1}{2}$	$\frac{1}{3}$	$\frac{1}{6} \times \frac{1}{6}$	$\frac{1}{2} \times \frac{1}{6}$	$\frac{1}{3} \times \frac{1}{6}$

Therefore, the mathematical expectation of his waiting time is

$$10 \times \frac{1}{2} + 30 \times \frac{1}{3} + 50 \times \frac{1}{36} + 70 \times \frac{1}{12} + 90 \times \frac{1}{18} \approx 27(\text{min}).$$

Part II Word Problems (14 marks for Question 9, 15 marks each for Questions 10 and 11, 44 marks in total)

9 Suppose that the line l: $y = kx + m$ (k, m are integers) intercepts an ellipse $\frac{x^2}{16} + \frac{y^2}{12} = 1$ at two different points A, B, and intercepts the hyperbola $\frac{x^2}{4} - \frac{y^2}{12} = 1$ at two different points C, D. Can the line l be such that $\overrightarrow{AC} + \overrightarrow{BD} = 0$? If yes, how many different possibilities are there for the line l? If no, explain the reason.

Solution For $\begin{cases} y = kx + m, \\ \frac{x^2}{16} + \frac{y^2}{12} = 1, \end{cases}$ by eliminating y and simplifying it, we get

$$(3 + 4k^2)x^2 + 8kmx + 4m^2 - 48 = 0.$$

Define $A(x_1, y_1)$, $B(x_2, y_2)$. Then $x_1 + x_2 = -\frac{8km}{3 + 4k^2}$.

$$\Delta_1 = (8km)^2 - 4(3 + 4k^2)(4m^2 - 48) > 0. \quad ①$$

For $\begin{cases} y = kx + m, \\ \frac{x^2}{4} - \frac{y^2}{12} = 1, \end{cases}$ by eliminating y and simplifying it, we get

$$(3 - k^2)x^2 - 2kmx - m^2 - 12 = 0.$$

Define $C(x_3, y_3)$, $D(x_4, y_4)$. Then $x_3 + x_4 = \frac{2km}{3 - k^2}$.

$$\Delta_2 = (-2km)^2 + 4(3 - k^2)(m^2 + 12) > 0. \quad ②$$

From $\overrightarrow{AC} + \overrightarrow{BD} = 0$ we get $(x_4 - x_2) + (x_3 - x_1) = 0$, which implies that $x_1 + x_2 = x_3 + x_4$.

Then

$$-\frac{8km}{3 + 4k^2} = \frac{2km}{3 - k^2}.$$

Therefore, $km = 0$ or $-\frac{4}{3 + 4k^2} = \frac{1}{3 - k^2}$ (discarded). Then a possible solution is either $k = 0$ or $m = 0$.

When $k = 0$, from ① and ② we have $-2\sqrt{3} < m < 2\sqrt{3}$. As m is an integer, it can be $-3, -2, -1, 0, 1, 2, 3$.

When $m = 0$, from ① and ② we have $-\sqrt{3} < k < \sqrt{3}$. As k is an integer, it can be $-1, 0, 1$.

Combining the results above, we conclude that there are nine lines in total satisfying the given conditions.

10 It is known that p, $q (q \neq 0)$ are real numbers; the equation $x^2 - px + q = 0$ has two real roots α, β; the sequence $\{a_n\}$ satisfies $a_1 = p$, $a_2 = p^2 - q$, $a_n = pa_{n-1} - qa_{n-2} (n = 3, 4, \ldots)$.

(1) Find the general expression of $\{a_n\}$ in terms of α, β.

(2) If $p = 1$, $q = \frac{1}{4}$, find the sum of the first n terms of $\{a_n\}$.

Solution I (1) By Vieta's theorem, we have $\alpha \times \beta = q \neq 0$, $\alpha + \beta = p$. Then

$$\begin{aligned} a_n &= pa_{n-1} - qa_{n-2} \\ &= (\alpha + \beta)a_{n-1} - \alpha\beta a_{n-2} (n = 3, 4, \ldots). \end{aligned}$$

This can be rewritten as

$$a_n - \beta a_{n-1} = \alpha(a_{n-1} - \beta a_{n-2}).$$

Let $b_n = a_{n+1} - \beta a_n$. Then $b_{n+1} = \alpha b_n (n = 1, 2, \ldots)$. This means that $\{b_n\}$ is a geometric sequence with the common ratio α. The first term of $\{b_n\}$ is

$$b_1 = a_2 - \beta a_1 = p^2 - q - \beta p = (\alpha + \beta)^2 - \alpha\beta - \beta(\alpha + \beta) = \alpha^2.$$

Therefore, $b_n = \alpha^2 \times \alpha^{n-1} = \alpha^{n+1}$. Then $a_{n+1} - \beta a_n = \alpha^{n+1}$. By rewriting

$$a_{n+1} = \alpha^{n+1} + \beta a_n (n = 1, 2, \ldots). \qquad ①$$

When $\Delta = p^2 - 4q = 0$, we have $\alpha = \beta \neq 0$, $a_1 = p = 2\alpha$. The expression ① becomes $a_{n+1} = \alpha^{n+1} + \alpha a_n$, i.e. $\frac{a_{n+1}}{\alpha^{n+1}} - \frac{a_n}{\alpha^n} = 1$.

Then $\left\{\frac{a_n}{\alpha^n}\right\}$ is an arithmetic sequence with the common difference 1, whose first term is $\frac{a_1}{\alpha} = \frac{2\alpha}{\alpha} = 2$. Therefore,

$$\frac{a_n}{\alpha^n} = 2 + 1 \times (n - 1) = n + 1.$$

As a result, the general expression of $\{a_n\}$ is

$$a_n = (n + 1)\alpha^n. \qquad ②$$

When $\Delta > 0$, $\alpha \neq \beta$, we have

$$\begin{aligned} a_{n+1} &= \alpha^{n+1} + \beta a_n \\ &= \beta a_n + \frac{\beta}{\beta - \alpha}\alpha^{n+1} - \frac{\alpha}{\beta - \alpha}\alpha^{n+1} (n = 1, 2, \ldots). \end{aligned}$$

By rewriting,

$$a_{n+1} + \frac{\alpha^{n+2}}{\beta - \alpha} = \beta\left(a_n + \frac{\alpha^{n+1}}{\beta - \alpha}\right)(n = 1, 2, \ldots).$$

Then $\left\{a_n + \frac{\alpha^{n+1}}{\beta - \alpha}\right\}$ becomes a geometric sequence with the common ratio β, whose first term is

$$a_1 + \frac{\alpha^2}{\beta - \alpha} = \alpha + \beta + \frac{\alpha^2}{\beta - \alpha} = \frac{\beta^2}{\beta - \alpha}.$$

Therefore,

$$a_n + \frac{\alpha^{n+1}}{\beta - \alpha} = \frac{\beta^2}{\beta - \alpha}\beta^{n-1}.$$

Then the general expression of $\{a_n\}$ is

$$a_n = \frac{\beta^{n+1} - \alpha^{n+1}}{\beta - \alpha} \ (n = 1, 2, \ldots). \quad ③$$

(2) Given $p = 1$, $q = \frac{1}{4}$, we have $\Delta = p^2 - 4q = 0$. Then $\alpha = \beta = \frac{1}{2}$. By the expression ②, the general expression of $\{a_n\}$ is

$$a_n = (n+1)\left(\frac{1}{2}\right)^n = \frac{n+1}{2^n} \ (n = 1, 2, \ldots).$$

Therefore, the sum of the first n terms of $\{a_n\}$ is

$$S_n = \frac{2}{2} + \frac{3}{2^2} + \cdots + \frac{n+1}{2^n}. \quad ④$$

Then

$$\frac{1}{2}S_n = \frac{2}{2^2} + \frac{3}{2^3} + \cdots + \frac{n+1}{2^{n+1}}. \quad ⑤$$

④−⑤,

$$\frac{1}{2}S_n = \frac{3}{2} - \frac{n+3}{2^{n+1}}.$$

We finally get

$$S_n = 3 - \frac{n+3}{2^n}.$$

Solution II (1) By Vieta's theorem, we have $\alpha \times \beta = q \neq 0$, $\alpha + \beta = p$. Then

$$a_1 = \alpha + \beta,\ a_2 = \alpha^2 + \alpha\beta + \beta^2. \qquad ⑥$$

The character equation of $\{a_n\}$ is $\lambda^2 - p\lambda + q = 0$, which has roots α, β. Then we can write down the general expression of $\{a_n\}$ according to the following different situations:

When $\alpha = \beta \neq 0$, $a_n = (A_1 + A_2 n)\alpha^n$. From ⑥, we have

$$(A_1 + A_2)\alpha = 2\alpha,$$
$$(A_1 + 2A_2)\alpha^2 = 3\alpha^2.$$

Then we get $A_1 = A_2 = 1$. Therefore,

$$a_n = (n+1)\alpha^n.$$

When $\alpha \neq \beta$, $a_n = A_1\alpha^n + A_2\beta^n$. From ⑥, we have

$$A_1\alpha + A_2\beta = \alpha + \beta,$$
$$A_1\alpha^2 + A_2\beta^2 = \alpha^2 + \alpha\beta + \beta^2.$$

Then we get $A_1 = \frac{-\alpha}{\beta - \alpha}$, $A_2 = \frac{\beta}{\beta - \alpha}$. Therefore,

$$a_n = \frac{-\alpha^{n+1}}{\beta - \alpha} + \frac{\beta^{n+1}}{\beta - \alpha} = \frac{\beta^{n+1} - \alpha^{n+1}}{\beta - \alpha}.$$

(2) The solution is the same as Solution I.

11 Find the maximum and minimum of the function

$$y = \sqrt{x + 27} + \sqrt{13 - x} + \sqrt{x}.$$

Solution The domain of y is $x \in [0, 13]$. We have

$$\begin{aligned} y &= \sqrt{x+27} + \sqrt{13-x} + \sqrt{x} \\ &= \sqrt{x+27} + \sqrt{13+2\sqrt{x(13-x)}} \\ &\geqslant \sqrt{27} + \sqrt{13} = 3\sqrt{3} + \sqrt{13}. \end{aligned}$$

The equality holds when $x = 0$. Therefore, the minimum of y is $3\sqrt{3} + \sqrt{13}$.

On the other hand, by the Cauchy inequality we have

$$\begin{aligned} y^2 &= (\sqrt{x} + \sqrt{x+27} + \sqrt{13-x})^2 \\ &\leqslant \left(\frac{1}{2} + 1 + \frac{1}{3}\right)[2x + (x+27) + 3(13-x)] \\ &= 121. \end{aligned}$$

The equality holds when $4x = 9(13-x) = x+27$. It is so for $x = 9$. Therefore, the maximum of y is 11.

China Mathematical Competition (Complementary Test)

2008 (Chongqing)

1 (50 marks) As shown in Fig. 1, $ABCD$ is a convex quadrilateral with $\angle B + \angle D < 180°$, and P is a moving point on the plane. Let

$$f(P) = PA \times BC + PD \times CA + PC \times AB.$$

(1) Prove that P, A, B, C are concyclic when $f(P)$ reaches the minimum.

(2) Suppose that point E is on the arc $\overset{\frown}{AB}$ of the circumscribed circle O

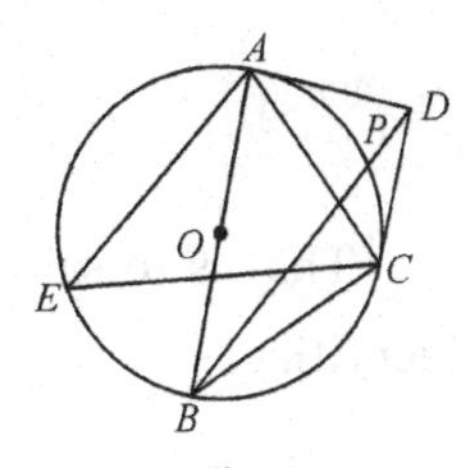

Fig. 1

of $\triangle ABC$, satisfying

$$\frac{AE}{AB}=\frac{\sqrt{3}}{2},\ \frac{BC}{EC}=\sqrt{3}-1,\ \angle ECB=\frac{1}{2}\angle ECA;$$

furthermore, DA, DC are tangent to $\odot O$, $AC=\sqrt{2}$. Find the minimum of $f(P)$.

Solution I (1) As shown is Fig. 1, by the Ptolemy inequality we have

$$PA\times BC+PC\times AB\geqslant PB\times AC.$$

Therefore,

$$\begin{aligned} f(P) &= PA\times BC+PC\times AB+PD\times CA \\ &\geqslant PB\times CA+PD\times CA \\ &= (PB+PD)\times CA. \end{aligned}$$

The equality holds if and only if P, A, B, C lie on $\odot O$ and P on $\overset{\frown}{AC}$. Furthermore, $PB+PD\geqslant BD$, and the equality holds if and only if P lies on line BD. Combining the results above, we have

$$f(P)_{\min}=BD\times CA.$$

This completes the proof that P, A, B, C are concyclic when $f(P)$ reaches the minimum.

(2) Denote $\angle ECB=\alpha$. Then $\angle ECA=2\alpha$. By the sine theorem we have

$$\frac{AE}{AB}=\frac{\sin 2\alpha}{\sin 3\alpha}=\frac{\sqrt{3}}{2}.$$

That is to say, $\sqrt{3}\sin 3\alpha=2\sin 2\alpha$. By using trigonometric identities,

$$\sqrt{3}(3\sin\alpha-4\sin^3\alpha)=4\sin\alpha\cos\alpha.$$

By simplification,

$$3\sqrt{3} - 4\sqrt{3}(1 - \cos^2\alpha) - 4\cos\alpha = 0.$$

That is to say, $4\sqrt{3}\cos^2\alpha - 4\cos\alpha - \sqrt{3} = 0.$

The solutions are $\cos\alpha = \frac{\sqrt{3}}{2}$ and $\cos\alpha = -\frac{1}{2\sqrt{3}}$ (discarded). Therefore $\alpha = 30^\circ$ and $\angle ECA = 60^\circ$.

On the other hand,

$$\frac{BC}{EC} = \sqrt{3} - 1 = \frac{\sin(\angle EAC - 30^\circ)}{\sin\angle EAC}.$$

That is to say

$$\frac{\sqrt{3}}{2}\sin\angle EAC - \frac{1}{2}\cos\angle EAC = (\sqrt{3} - 1)\sin\angle EAC.$$

Then

$$\frac{2 - \sqrt{3}}{2}\sin\angle EAC = \frac{1}{2}\cos\angle EAC.$$

Therefore,

$$\tan\angle EAC = \frac{1}{2 - \sqrt{3}} = 2 + \sqrt{3}.$$

We obtain $\angle EAC = 75^\circ$ and $\angle AEC = 45^\circ$.

Since $\triangle ADC$ is an isosceles triangle and $AC = \sqrt{2}$, we have $CD = 1$. Furthermore, $\triangle ABC$ is an isosceles triangle, so $BC = \sqrt{2}$ and $AB = 2$. Then

$$BD^2 = AB^2 + AD^2 = 4 + 1 = 5.$$

We have $BD = \sqrt{5}$. Therefore,

$$f(P)_{\min} = BD \times CA = \sqrt{5} \times \sqrt{2} = \sqrt{10}.$$

Solution II (1) As shown in Fig. 2, the line BD intercepts the circumscribed circle O of $\triangle ABC$ at point P_0, which lies on BD because D is outside of $\odot O$. Through A, C, D draw lines perpendicular to P_0A, P_0C, P_0D, respectively, and they constitute $\triangle A_1B_1C_1$ by intercepting with each other. It is easy to see that P_0 lies in $\triangle ABC$, as well as in $\triangle A_1B_1C_1$. Denoting the three inner angles of $\triangle ACD$ by x, y, z, respectively, we have

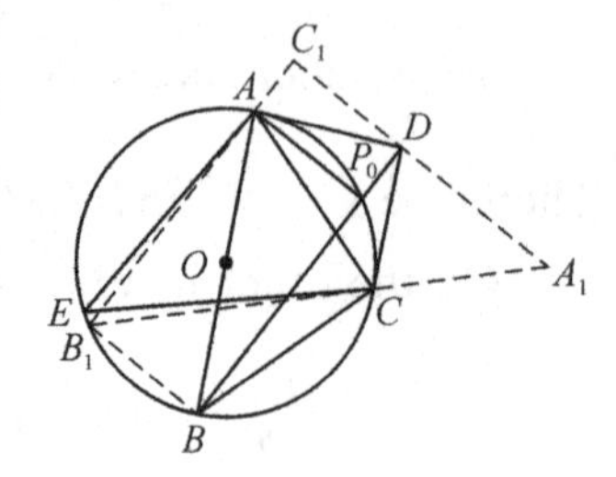

Fig. 2

$$\angle AP_0C = 180^\circ - y = z + x.$$

Furthermore, from $B_1C_1 \perp P_0A$ and $B_1A_1 \perp P_0C$, we have $\angle B_1 = y$. In a similar way, we have $\angle A_1 = x$ and $\angle C_1 = z$. It follows that $\triangle A_1B_1C_1 \backsim \triangle ABC$. Now let

$$B_1C_1 = \lambda BC,\ C_1A_1 = \lambda CA,\ A_1B_1 = \lambda AB.$$

Then for any point M on the plane, we have

$$\begin{aligned}\lambda f(P_0) &= \lambda(P_0A \times BC + P_0D \times CA + P_0C \times AB)\\ &= P_0A \times B_1C_1 + P_0D \times C_1A_1 + P_0C \times A_1B_1\\ &= 2S_{\triangle A_1B_1C_1}\\ &\leqslant MA \times B_1C_1 + MD \times C_1A_1 + MC \times A_1B_1\\ &= \lambda(MA \times BC + MD \times CA + MC \times AB)\\ &= \lambda f(M).\end{aligned}$$

That is to say, $f(P_0) \leqslant f(M)$. Since M is an arbitrary point, we know that $f(P)$ reaches the minimum at P_0, and at the same time P_0, A, B, C are concyclic. This completes the proof.

(2) From (1) we know that the minimum of $f(P)$ is

$$f(P_0) = \frac{2}{\lambda} S_{\triangle A_1B_1C_1} = 2\lambda S_{\triangle ABC}.$$

In the same way as is shown in the proof of (2) in Solution Ⅰ, we find that both $\triangle ADC$ and $\triangle ABC$ are right-angled isosceles triangles; so we have $CD = AD = \frac{AC}{\sqrt{2}} = 1$, $AB = \sqrt{2}AC = 2$, $S_{\triangle ABC} = 1$,

$$BD = \sqrt{AB^2 + AD^2} = \sqrt{5}.$$

Furthermore, since $\triangle A_1B_1C_1 \backsim \triangle ABC$, $\angle AB_1B = \angle AB_1C + \angle BB_1C = 90°$, AB_1BD is a rectangle. It follows that $B_1C_1 = BD = \sqrt{5}$. Therefore $\lambda = \frac{\sqrt{5}}{\sqrt{2}}$ and

$$f(P)_{\min} = 2 \times \frac{\sqrt{5}}{\sqrt{2}} \times 1 = \sqrt{10}.$$

Solution Ⅲ (1) We discuss the problem on the complex plane, and regard points A, B, C as complex numbers. Then by the triangle inequality we have

$$|\overrightarrow{PA} \cdot \overrightarrow{BC}| + |\overrightarrow{PC} \cdot \overrightarrow{AB}| \geqslant |\overrightarrow{PA} \cdot \overrightarrow{BC} + \overrightarrow{PC} \cdot \overrightarrow{AB}|.$$

That is to say,

$$\begin{aligned}
&|(A-P)(C-B)| + |(C-P)(B-A)| \\
\geqslant &|(A-P)(C-B) + (C-P)(B-A)| \\
= &|-P \times C - A \times B + C \times B + P \times A| \\
= &|(B-P)(C-A)| \\
= &|\overrightarrow{PB} \cdot \overrightarrow{AC}|. \qquad ①
\end{aligned}$$

Then

$$\begin{aligned}
&|\overrightarrow{PA}\cdot\overrightarrow{BC}|+|\overrightarrow{PC}\cdot\overrightarrow{AB}|+|\overrightarrow{PD}\cdot\overrightarrow{AC}|\\
\geqslant&|\overrightarrow{PB}\cdot\overrightarrow{AC}|+|\overrightarrow{PD}\cdot\overrightarrow{AC}|\\
=&(\overrightarrow{PB}+\overrightarrow{PD})\cdot|\overrightarrow{AC}|\\
=&|\overrightarrow{BD}|\cdot|\overrightarrow{AC}|. \qquad ②
\end{aligned}$$

The equality in ① holds only when the complex numbers $(A-P)(C-B)$ and $(C-P)(B-A)$ are in the same direction. This means that there is a real $\lambda > 0$ such that

$$(A-P)(C-B)=\lambda(C-P)(B-A).$$

That is to say,

$$\frac{A-P}{C-P}=\lambda\frac{B-A}{C-B}.$$

Therefore,

$$\arg\left(\frac{A-P}{C-P}\right)=\arg\left(\frac{B-A}{C-B}\right),$$

which means that the angle of rotation from $\overrightarrow{PC}$ to $\overrightarrow{PA}$ is equal to that from $\overrightarrow{BC}$ to $\overrightarrow{AB}$. It follows that P, A, B, C are concyclic.

The equality in ② holds only when B, P, D are collinear and P lies on the segment BD. This means that $f(P)$ reaches the minimum when P lies on the circumscribed circle of $\triangle ABC$ and P, A, B, C are concyclic.

(2) From (1) we know that $f(P)_{\min} = BD \times CA$. The following steps are the same as those in the proof of (2) in Solution Ⅰ.

2 (50 marks) Let $f(x)$ be a periodic function with periods T and 1, satisfying $0 < T < 1$. Prove that:

(1) If T is a rational number, there exists a prime p such

that $\frac{1}{p}$ is also a period of $f(x)$;

(2) If T is irrational, there exists a sequence of irrational numbers $\{a_n\}$, satisfying $1 > a_n > a_{n+1} > 0(n = 1, 2, \ldots)$ and each a_n is a period of $f(x)$.

Solution (1) T is a rational number, and there exist positive integers m, n such that $T = \frac{n}{m}$ and $(m, n) = 1$. There are two integers a, b such that $ma + nb = 1$. It follows that

$$\frac{1}{m} = \frac{ma + nb}{m} = a \times 1 + b \times T$$

is also a period of $f(x)$. Furthermore, $m \geqslant 2$ because $0 < T < 1$. Then $m = pm'$, p being a prime. Therefore, $\frac{1}{p} = m' \times \frac{1}{m}$ is a period of $f(x)$.

(2) If T is irrational, we define $a_1 = 1 - \left[\frac{1}{T}\right] \times T$. Then $0 < a_1 < 1$ and a_1 is irrational. Define recursively

$$a_{n+1} = 1 - \left[\frac{1}{a_n}\right] \times a_n.$$

It is easy to know by induction that each a_n is irrational and $0 < a_n < 1$. On the other hand, we have $\frac{1}{a_n} - \left[\frac{1}{a_n}\right] < 1$. That is to say,

$$1 < a_n + \left[\frac{1}{a_n}\right] \times a_n.$$

Then

$$a_{n+1} = 1 - \left[\frac{1}{a_n}\right] \times a_n < a_n.$$

This means that $\{a_n\}$ is a decreasing sequence.

Finally, it is easy to prove by the definition of $\{a_n\}$ and induction that each a_n is a period of $f(x)$.

3 (50 marks) Suppose that $a_k > 0$, $k = 1, 2, \ldots, 2008$. Prove that if and only if $\sum_{k=1}^{2008} a_k > 1$, there is a sequence $\{x_n\}$ satisfying

(1) $0 = x_0 < x_n < x_{n+1}$, $n = 1, 2, 3, \ldots$;

(2) $\lim_{n\to\infty} x_n$ exists;

(3) $x_n - x_{n-1} = \sum_{k=1}^{2008} a_k x_{n+k} - \sum_{k=0}^{2007} a_{k+1} x_{n+k}$, $n = 1, 2, 3, \ldots$.

Solution Proof of necessity: Assume that there exists $\{x_n\}$ satisfying (1)-(3). Notice that the expression in (3) can be written as

$$x_n - x_{n-1} = \sum_{k=1}^{2008} a_k (x_{n+k} - x_{n+k-1}),\ n \in \mathbf{N}.$$

As $x_0 = 0$, we then have

$$\begin{aligned} x_n &= \sum_{l=1}^{n} (x_l - x_{l-1}) = \sum_{l=1}^{n} \sum_{k=1}^{2008} a_k (x_{l+k} - x_{l+k-1}) \\ &= \sum_{k=1}^{2008} \sum_{l=1}^{n} a_k (x_{l+k} - x_{l+k-1}) \\ &= \sum_{k=1}^{2008} a_k (x_{n+k} - x_k). \end{aligned}$$

From (2) we are able to define $b = \lim_{n\to\infty} x_n$. Let $n \to \infty$ in the above expression. Then we have

$$\begin{aligned} b &= \sum_{k=1}^{2008} a_k (b - x_k) = b \sum_{k=1}^{2008} a_k - \sum_{k=1}^{2008} a_k x_k \\ &< b \sum_{k=1}^{2008} a_k. \end{aligned}$$

Therefore, $\sum_{k=1}^{2008} a_k > 1$.

Proof of sufficiency: Assume that $\sum_{k=1}^{2008} a_k > 1$. Define a polynomial function by

$$f(s) = -1 + \sum_{k=1}^{2008} a_k s^k,\ s \in [0, 1].$$

$f(s)$ is strictly increasing on the interval $[0, 1]$. In the meantime,

$$f(0) = -1 < 0,\ f(1) = -1 + \sum_{k=1}^{2008} a_k > 0,$$

and there then exists a unique $0 < s_0 < 1$ such that $f(s_0) = 0$. Now, define $x_n = \sum_{k=1}^{n} s_0^k$, $n \in \mathbf{N}$. It is easy to see that $\{x_n\}$ satisfies (1) and

$$x_n = \sum_{k=1}^{n} s_0^k = \frac{s_0 - s_0^{n+1}}{1 - s_0}.$$

On the other hand, $\lim_{n\to\infty} s_0^{n+1} = 0$ as $0 < s_0 < 1$. Then we have

$$\lim_{n\to\infty} x_n = \lim_{n\to\infty} \frac{s_0 - s_0^{n+1}}{1 - s_0} = \frac{s_0}{1 - s_0}.$$

This means that $\{x_n\}$ satisfies (2). Finally, we have

$$0 = f(s_0) = -1 + \sum_{k=1}^{2008} a_k s_0^k.$$

That is to say, $\sum_{k=1}^{2008} a_k s_0^k = 1$. Then we have

$$\begin{aligned} x_n - x_{n-1} &= s_0^n \\ &= \left(\sum_{k=1}^{2008} a_k s_0^k\right) s_0^n = \sum_{k=1}^{2008} a_k s_0^{n+k} \\ &= \sum_{k=1}^{2008} a_k (x_{n+k} - x_{n+k-1}). \end{aligned}$$

Therefore, $\{x_n\}$ also satisfies (3). This completes the proof.

2009 (Heilongjiang)

1 (50 marks) As shown in the Fig. 1, M, N are the midpoints of arcs $\overparen{BC}$, $\overparen{AC}$ respectively, which are on the circumscribed circle of an acute triangle $\triangle ABC$ ($\angle A < \angle B$). Through point C draw $PC \parallel MN$, intercepting the circle Γ at point P. I is the inner center of $\triangle ABC$. Extend line PI to intercept Γ at point T.

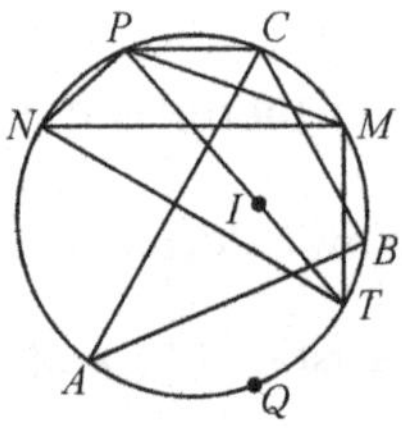

Fig. 1

(1) Prove that $MP \times MT = NP \times NT$;

(2) For an arbitrary point $Q(\neq A, T, B)$ on arc $\overparen{AB}$ (not containing C), denote the inner centers of $\triangle AQC$, $\triangle QCB$ by I_1, I_2, respectively. Prove that Q, I_1, I_2, T are concyclic.

Solution (1) As shown in Fig. 2, join NI, MI. Since $PC \parallel MN$ and P, C, M, N are concyclic, $PCMN$ is an isosceles trapezoid. Therefore, $NP = MC$, $PM = NC$.

Join AM, CI. Then AM intercepts CI at I. We have

$$\begin{aligned}\angle MIC &= \angle MAC + \angle ACI \\ &= \angle MCB + \angle BCI \\ &= \angle MCI.\end{aligned}$$

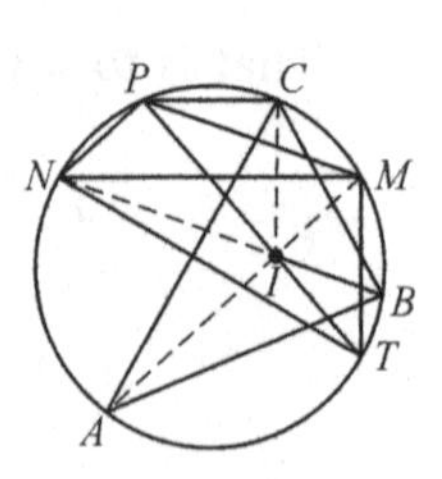

Fig. 2

Therefore, $MC = MI$. In the same way, $NC = NI$. Then $NP = MI$, $PM = NI$.

This means that $MPNI$ is a parallelogram. Therefore, $S_{\triangle PMT} = S_{\triangle PNT}$, for the two triangles having the same base and height.

On the other hand, $\angle TNP + \angle PMT = 180°$, as P, N, T, M are concyclic. Then we have

$$\begin{aligned} S_{\triangle PMT} &= \frac{1}{2}PM \times MT\sin\angle PMT \\ &= S_{\triangle PNT} = \frac{1}{2}PN \times NT\sin\angle PNT \\ &= \frac{1}{2}PN \times NT\sin\angle PMT. \end{aligned}$$

Therefore, $MP \times MT = NP \times NT$.

(2) As shown in Fig. 3, we have

$$\angle NCI_1 = \angle NCA + \angle ACI_1 = \angle NQC + \angle QCI_1 = \angle CI_1N.$$

Therefore, $NC = NI_1$. In the same way, $MC = MI_2$.

From $MP \times MT = NP \times NT$ we get

$$\frac{NT}{MP} = \frac{MT}{NP}.$$

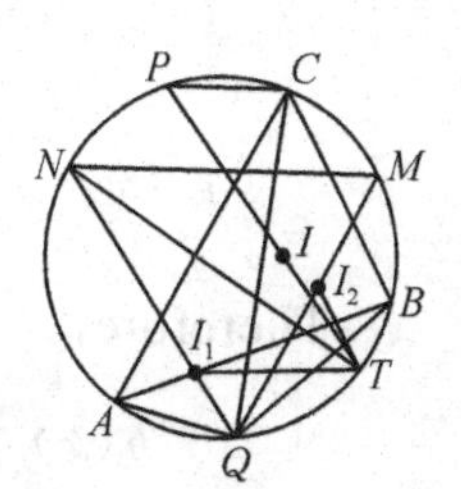

Fig. 3

From (1) we know that $MP = NC$, $NP = MC$. Then

$$\frac{NT}{NI_1} = \frac{MT}{MI_2}.$$

Furthermore,

$$\angle I_1NT = \angle QNT = \angle QMT = \angle I_2MT.$$

Therefore, $\triangle I_1NT \backsim \triangle I_2MT$. Consequently, $\angle NTI_1 =$

$\angle MTI_2$. Then we have

$$\angle I_1QI_2 = \angle NQM = \angle NTM = \angle I_1TI_2.$$

This means that Q, I_1, I_2, T are concyclic.

2 (50 marks) Prove that

$$-1 < \left(\sum_{k=1}^{n} \frac{k}{k^2+1}\right) - \ln n \leqslant \frac{1}{2}, \quad n = 1, 2, \ldots.$$

Solution We first prove that

$$\frac{x}{1+x} < \ln(1+x) < x, \ x > 0. \tag{1}$$

Let

$$h(x) = x - \ln(1+x),$$

$$g(x) = \ln(1+x) - \frac{x}{1+x}.$$

Then, for $x > 0$,

$$h'(x) = 1 - \frac{1}{1+x} > 0,$$

$$g'(x) = \frac{1}{1+x} - \frac{1}{(1+x)^2} = \frac{x}{(1+x)^2} > 0.$$

Therefore,

$$h(x) > h(0) = 0, \ g(x) > g(0) = 0.$$

This completes the proof of the inequalities ①.

Now let $x = \frac{1}{n}$ in ①. We have

$$\frac{1}{n+1} < \ln\left(1 + \frac{1}{n}\right) < \frac{1}{n}. \tag{2}$$

Let

$$x_n = \sum_{k=1}^{n} \frac{k}{k^2+1} - \ln n.$$

Then

$$\begin{aligned} x_n - x_{n-1} &= \frac{n}{n^2+1} - \ln\left(1+\frac{1}{n-1}\right) \\ &< \frac{n}{n^2+1} - \frac{1}{n} \\ &= -\frac{1}{n(n^2+1)} < 0. \end{aligned}$$

Therefore, $x_n < x_{n-1} < \cdots < x_1 = \frac{1}{2}$.

Furthermore,

$$\begin{aligned} \ln n &= (\ln n - \ln(n-1)) + (\ln(n-1) - \ln(n-2)) \\ &\quad + \cdots + (\ln 2 - \ln 1) + \ln 1 \\ &= \sum_{k=1}^{n-1} \ln\left(1+\frac{1}{k}\right). \end{aligned}$$

Consequently,

$$\begin{aligned} x_n &= \sum_{k=1}^{n} \frac{k}{k^2+1} - \sum_{k=1}^{n-1} \ln\left(1+\frac{1}{k}\right) \\ &= \sum_{k=1}^{n-1}\left(\frac{k}{k^2+1} - \ln\left(1+\frac{1}{k}\right)\right) + \frac{n}{n^2+1} \\ &> \sum_{k=1}^{n-1}\left(\frac{k}{k^2+1} - \frac{1}{k}\right) \\ &= -\sum_{k=1}^{n-1} \frac{1}{(k^2+1)k} \\ &\geqslant -\sum_{k=1}^{n-1} \frac{1}{(k+1)k} \\ &= -1 + \frac{1}{n} > -1. \end{aligned}$$

This completes the proof.

3 (50 marks) Suppose that k, l are two positive integers.

Prove that:

There are infinite many positive integers $m \geqslant k$ such that $\binom{m}{k}$ and l are relatively prime.

Solution I Let $m = k + t \times l \times (k!)$, where t is any positive integer. To prove that $\left(\binom{m}{k}, l\right) = 1$, we only need to prove that for any prime factor p of l, $p \nmid \binom{m}{k}$.

If $p \nmid k!$, we have

$$
\begin{aligned}
k!\binom{m}{k} &= \prod_{i=1}^{k}(m-k+i) \\
&= \prod_{i=1}^{k}[i + tl(k!)] \\
&\equiv \prod_{i=1}^{k} i \equiv k! \pmod{p}.
\end{aligned}
$$

Therefore, $p \nmid \binom{m}{k}$.

If $p \mid k!$, there exists integer $\alpha \geqslant 1$ such that $p^{\alpha} \mid k!$ but $p^{\alpha+1} \nmid k!$. Then $p^{\alpha+1} \mid l(k!)$.

We have

$$
\begin{aligned}
k!\binom{m}{k} &= \prod_{i=1}^{k}(m-k+i) \\
&= \prod_{i=1}^{k}[i + tl(k!)] \\
&\equiv \prod_{i=1}^{k} i \equiv k! \pmod{p^{\alpha+1}}.
\end{aligned}
$$

Therefore, $p^{\alpha} \mid k!\binom{m}{k}$ and $p^{\alpha+1} \nmid k!\binom{m}{k}$. Since $p^{\alpha} \mid k!$, we get $p \nmid \binom{m}{k}$. The proof is completed.

Solution II Let $m = k + t \times l \times (k!)^2$, where t is any positive integer. The following proof steps are similar to those in Solution I, and are omitted.

4 (50 marks) Suppose that a matrix of nonnegative entries,

$$P = \begin{bmatrix} x_{11} & x_{12} & x_{13} & x_{14} & x_{15} & x_{16} & x_{17} & x_{18} & x_{19} \\ x_{21} & x_{22} & x_{23} & x_{24} & x_{25} & x_{26} & x_{27} & x_{28} & x_{29} \\ x_{31} & x_{32} & x_{33} & x_{34} & x_{35} & x_{36} & x_{37} & x_{38} & x_{39} \end{bmatrix}$$

has the following properties:

(1) Numbers in a row are different from each other;

(2) The sum of the numbers in a column from the first six columns is 1;

(3) $x_{17} = x_{28} = x_{39} = 0$;

(4) x_{27}, x_{37}, x_{18}, x_{38}, x_{19}, $x_{29} > 1$.

Assume that matrix S is constituted by the first three columns of P, i.e.

$$S = \begin{bmatrix} x_{11} & x_{12} & x_{13} \\ x_{21} & x_{22} & x_{23} \\ x_{31} & x_{32} & x_{33} \end{bmatrix}$$

has the following property:

(O) For any column $\begin{bmatrix} x_{1k} \\ x_{2k} \\ x_{3k} \end{bmatrix}$ $(k = 1, 2, \ldots, 9)$ in P, there exists $i \in \{1, 2, 3\}$ such that

$$x_{ik} \leqslant u_i = \min\{x_{i1}, x_{i2}, x_{i3}\}. \quad ①$$

Prove that:

(i) For different $i\ (=1, 2, 3)$, $u_i = \min\{x_{i1}, x_{i2}, x_{i3}\}$ comes from different columns in S;

(ii) There exists a unique column $\begin{bmatrix} x_{1k^*} \\ x_{2k^*} \\ x_{3k^*} \end{bmatrix}$ $(k^* \neq 1, 2, 3)$ in P such that the matrix

$$S' = \begin{bmatrix} x_{11} & x_{12} & x_{1k^*} \\ x_{21} & x_{22} & x_{2k^*} \\ x_{31} & x_{32} & x_{3k^*} \end{bmatrix}$$

also has property (O).

Solution Proof of (i). Assume that it is not true. There is a column in S which contains no u_i. We may say that $u_i \neq x_{i2}$, $i = 1, 2, 3$. By property (1), we have $u_i < x_{i2}$, $i = 1, 2, 3$. On the other hand, let $k = 2$ in ①. Then by property (O), there exists $i_0 \in \{1, 2, 3\}$ such that $x_{i_0 2} \leqslant u_i$. The contradiction means the assumption is not valid. This completes the proof of (i).

Proof of (ii). By the drawer principle, we know that at least two of the three numbers

$$\min\{x_{11}, x_{12}\},\ \min\{x_{21}, x_{22}\},\ \min\{x_{31}, x_{32}\}$$

are in the same column. We may say that

$$\min\{x_{21}, x_{22}\} = x_{22},\ \min\{x_{31}, x_{32}\} = x_{32}.$$

By (i), we know that the first column of S contains a u_i, and it must be $u_1 = x_{11}$; the second column also contains a u_i, assuming that it is $u_2 = x_{22}$, and then it must be $u_3 = x_{33}$.

Define $M = \{1, 2, \ldots, 9\}$ and

$$I = \{k \in M \mid x_{ik} > \min\{x_{i1}, x_{i2}\},\ i = 1, 3\}.$$

Obviously, $I = \{k \in M \mid x_{1k} > x_{11}, x_{3k} > x_{32}\}$, and 1, 2, $3 \notin I$. Since $x_{18}, x_{38} > 1 \geqslant x_{11}, x_{32}$, we have $8 \in I$. Therefore, $I \neq \varnothing$. Consequently, $\exists k^* \in I$ such that $x_{2k^*} = \max\{x_{2k} \mid k \in I\}$. Of course, $k^* \neq 1, 2, 3$.

We now prove that

$$S' = \begin{bmatrix} x_{11} & x_{12} & x_{1k^*} \\ x_{21} & x_{22} & x_{2k^*} \\ x_{31} & x_{32} & x_{3k^*} \end{bmatrix}$$

has property (O).

By the definition of I, we know that

$$x_{1k^*} > x_{11} = u_1, \ x_{3k^*} > x_{32} \geqslant u_3.$$

Let $k^* = k$ in (ⅰ), and we get $x_{2k^*} \leqslant u_2$ according to property (O) of S. Then define

$$u_1' = u_1, \ u_2' = \min\{x_{21}, x_{22}, x_{2k^*}\} = x_{2k^*}, \ u_3' = u_3.$$

We claim that, for any $k \in M$, there exists $i \in \{1, 2, 3\}$ such that $u_i' \geqslant x_{ik}$. Otherwise, we would have $x_{ik} > \min\{x_{i1}, x_{i2}\}$, $i = 1, 3$ and $x_{2k} > x_{2k^*}$, contradicting the definition of k^*.

Therefore, S' has property (O).

Secondly, we prove the uniqueness of S'. Assume that $\exists k_0 \in M$ such that

$$\hat{S} = \begin{bmatrix} x_{11} & x_{12} & x_{1k_0} \\ x_{21} & x_{22} & x_{2k_0} \\ x_{31} & x_{32} & x_{3k_0} \end{bmatrix}$$

also has property (O). Without loss of generality, we assume that

$$u_i = \min\{x_{i1}, x_{i2}, x_{i3}\} = x_{ii}, i = 1, 2, 3, \quad ②$$

$$x_{32} < x_{31}.$$

Since $x_{32} < x_{31}$, $x_{22} < x_{21}$, and by (i), we have

$$\hat{u}_1 = \min\{x_{11}, x_{12}, x_{1k_0}\} = x_{11}.$$

By (i) again, we have either

(a) $\hat{u}_3 = \min\{x_{31}, x_{32}, x_{3k_0}\} = x_{3k_0}$ or

(b) $\hat{u}_2 = \min\{x_{21}, x_{22}, x_{2k_0}\} = x_{2k_0}$.

If (a) is true, we will have

$$\hat{u}_1 = x_{11}, \hat{u}_2 = x_{22}, \hat{u}_3 = x_{3k_0}. \quad ③$$

For $3 \in M$, since both $\hat{S}$ and S have property (O), we will get

$$x_{33} \leqslant \hat{u}_3 = x_{3k_0}, x_{3k_0} \leqslant u_3 = x_{33}.$$

By property (1) of P, we have $3 = k_0$. Therefore, $\hat{S} = S$.

If (b) is true, we will have

$$\hat{u}_1 = x_{11}, \hat{u}_2 = x_{2k_0}, \hat{u}_3 = x_{32}. \quad ④$$

Since $\hat{S}$ has property (O), we know that, for $k^* \in M$, there exists $i \in \{1, 2, 3\}$ such that $\hat{u}_i \geqslant x_{ik}$. As $k^* \in I$, and by ②, ④, it must be $x_{2k^*} \leqslant \hat{u}_2 = x_{2k}$.

On the other hand, S also has property (O). Then, in a similar way, we get $x_{2k} \leqslant u_2' = x_{2k^*}$.

Therefore, $k^* = k$. This completes the proof.

Chinese Mathematical Olympiad

2009 (Qionghai, Hainan)

First Day

(0800 - 1230; January 9, 2009)

1. Given an acute triangle PBC, $PB \neq PC$. Let points A, D be on sides PB and PC, respectively. Let M, N be the midpoints of segments BC and AD, respectively. Lines AC and BD intersect at point O. Draw $OE \perp AB$ at point E and $OF \perp CD$ at point F.

(1) Prove that if A, B, C, D are concyclic, then

$$EM \times FN = EN \times FM.$$

(2) Are the four points A, B, C, D always concyclic if $EM \times FN = EN \times FM$? Prove your answer. (Posed by Xiong Bin)

Proof (1) Denote by Q, R the midpoints of OB, OC, respectively. It is easy to see that

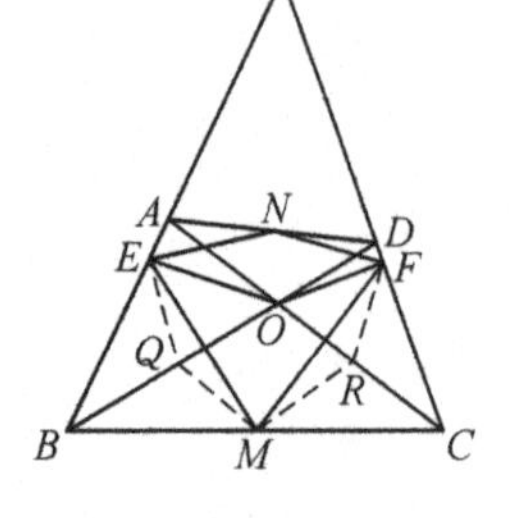

$$EQ = \frac{1}{2}OB = RM, \ MQ = \frac{1}{2}OC = RF,$$

and

$$\angle EQM = \angle EQO + \angle OQM = 2\angle EBO + \angle OQM,$$
$$\angle MRF = \angle FRO + \angle ORM = 2\angle FCO + \angle ORM.$$

Because of that A, B, C, D are concyclic, and Q, R are the midpoints of OB, OC, and we have

$$\angle EBO = \angle FCO, \ \angle OQM = \angle ORM.$$

So $\angle EQM = \angle MRF$, which implies that $\triangle EQM \cong \triangle MRF$, and $EM = FM$.

Similarly, we have $EN = FN$, so $EM \times FN = EN \times FM$ holds.

(2) Suppose that $OA = 2a$, $OB = 2b$, $OC = 2c$, $OD = 2d$ and

$$\angle OAB = \alpha, \ \angle OBA = \beta, \ \angle ODC = \gamma, \ \angle OCD = \theta.$$

Then

$$\begin{aligned}\cos\angle EQM &= \cos(\angle EQO + \angle OQM)\\ &= \cos(2\beta + \angle AOB)\\ &= -\cos(\alpha - \beta).\end{aligned}$$

So

$$\begin{aligned} EM^2 &= EQ^2 + QM^2 - 2EQ \times QM \times \cos\angle EQM \\ &= b^2 + c^2 + 2bc\cos(\alpha - \beta). \end{aligned}$$

Making similar equations for EN, FN, FM, we have

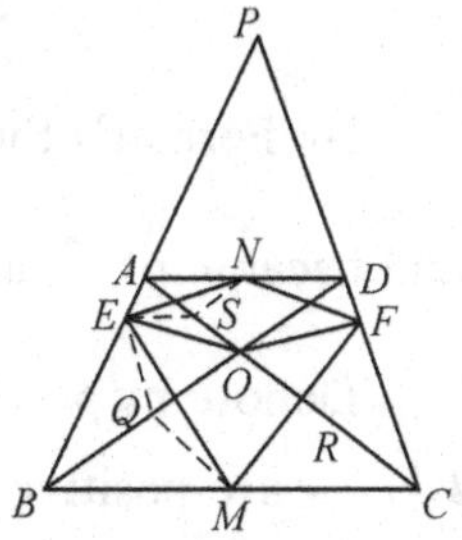

$$\begin{aligned} & EN \times FM = EM \times FN \\ \Leftrightarrow\ & EN^2 \times FM^2 = EM^2 \times FN^2 \\ \Leftrightarrow\ & (a^2 + d^2 + 2ad\cos(\alpha - \beta)) \times (b^2 + c^2 + 2bc\cos(\gamma - \theta)) = (a^2 + d^2 + 2ad\cos(\gamma - \theta)) \times (b^2 + c^2 + 2bc\cos(\alpha - \beta)) \\ \Leftrightarrow\ & (\cos(\gamma - \theta) - \cos(\alpha - \beta))(ab - cd)(ac - bd) = 0. \end{aligned}$$

Because $\alpha + \beta = \gamma + \theta$, $\cos(\gamma - \theta) - \cos(\alpha - \beta) = 0$ holds if and only if $\alpha = \gamma$, $\beta = \theta$ (i.e. A, B, C, D are concyclic) or $\alpha = \theta$, $\beta = \gamma$ (which follows $AB \parallel CD$, a contradiction.); $ab - cd = 0$ holds if and only if $AD \parallel BC$; $ac - bd = 0$ holds if and only if A, B, C, D are concyclic.

So, when $AD \parallel BC$ holds, we also have

$$EM \times FN = EN \times FM.$$

We know that A, B, C, D are not concyclic in this case because $PB \neq PC$, so the answer is "false".

2 Find all pairs (p, q) of prime numbers such that $pq \mid 5^p + 5^q$. (Posed by Fu Yunhao)

Solution If $2 \mid pq$, we suppose that $p = 2$ without loss of generality, and then $q \mid 5^q + 25$. By Fermat's theorem we have $q \mid 5^q - 5$, so $q \mid 30$, where $(2, 3)$ and $(2, 5)$ are solutions [$(2, 2)$ does not fit].

If $5 \mid pq$, we suppose that $p = 5$ without loss of generality and then $5q \mid 5^q + 5^5$. By Fermat's theorem we have $q \mid 5^{q-1} - 1$, so $q \mid 313$, where (5, 5) and (5, 313) are solutions.

Otherwise, we have $pq \mid 5^{p-1} + 5^{q-1}$, and so

$$5^{p-1} + 5^{q-1} \equiv 0 \pmod{p}. \qquad ①$$

By Fermat's theorem, $5^{p-1} \equiv 1 \pmod{p}$, ②

and because of ① and ②, $5^{q-1} \equiv -1 \pmod{p}$. ③

Denote by $p-1 = 2^k(2r-1)$, $q-1 = 2^l(2s-1)$, where k, l, r, s are positive integers.

If $k \leqslant l$, because of ② and ③ we get

$$\begin{aligned} 1 &= 1^{2^{l-k}(2s-1)} \equiv (5^{p-1})^{2^{l-k}(2s-1)} \\ &= 5^{2^l(2r-1)(2s-1)} = (5^{q-1})^{2r-1} \equiv (-1)^{2r-1} \\ &\equiv -1 \pmod{p}, \end{aligned}$$

a contradiction of $p \neq 2$. So $k > l$.

But we have $k < l$ by a similar argument — a contradiction. Therefore, all possible pairs of primes (p, q) are (2, 3), (3, 2), (2, 5), (5, 2), (5, 5), (5, 313) and (313, 5).

3 Let m, n be integers with $4 < m < n$, and $A_1A_2 \ldots A_{2n+1}$ be a regular $2n+1$ polygon. In addition, let $P = \{A_1, A_2, \ldots, A_{2n+1}\}$. Find the number of convex m-gons with exactly two acute internal angles whose vertices are all in P. (Posed by Leng Gangsong)

Solution Notice that if a convex m-gon whose vertex set is contained in P has exactly two acute angles, they must be at consecutive vertices: for otherwise there would be two disjoint pairs of sides that take up more than half of the circle each. Now assume that the last vertex, clockwise, of these four vertices that make up two acute angles is fixed; this reduces the total number of regular m-gons $2n+1$ times and we will later multiply by this factor.

Suppose that the larger arc that the first and the last of these four vertices make contains k points, and the other arc contains $2n-1-k$ points. For each k, the vertices of the m-polygon on the smaller arc may be arranged in $\binom{2n-1-k}{m-4}$ ways, and the two vertices on the larger arc may be arranged in $(k-n-1)^2$ ways (so that the two angles cut off more than half of the circle).

The total number of polygons given by k is thus $(k-n-1)^2 \times \binom{2n-1-k}{m-4}$. Summation over all k and a change of variable shows that the total number of polygons (divided by a factor of $2n+1$) is $\sum_{k\geqslant 0} k^2 \times \binom{n-k-2}{m-4}$.

This can be proven to be exactly $\binom{n}{m-1}+\binom{n+1}{m-1}$ by double induction on $n>m$ and $m>4$. The base cases $n=m+1$ and $m=5$ are readily calculated. The induction step is

$$\begin{aligned}
&\sum_{k\geqslant 0} k^2 \times \binom{n-k-2}{m-4} \\
&= \sum_{k\geqslant 0} k^2 \times \binom{(n-1)-k-2}{m-4} + \sum_{k\geqslant 0} k^2 \times \binom{(n-1)-k-2}{(m-1)-4} \\
&= \binom{n-1}{m-1} + \binom{n}{m-1} + \binom{n-1}{m-2} + \binom{n}{m-2} \\
&= \binom{n}{m-1} + \binom{n+1}{m-1}.
\end{aligned}$$

So the total number of $(2n+1)$-gons is

$$(2n+1)\left(\binom{n}{m-1}+\binom{n+1}{m-1}\right).$$

Second Day

(0800 - 1230; January 10, 2009)

4 Let $n \geqslant 3$ be a given integer, and $a_1, a_2, \ldots, a_n$ be real numbers satisfying $\min\limits_{1\leqslant i<j\leqslant n} |a_i - a_j| = 1$. Find the minimum value of $\sum\limits_{k=1}^{n} |a_k|^3$. (Posed by Zhu Huawei)

Solution Without loss of generality, we may assume that $a_1 < a_2 < \cdots < a_n$, and note also that

$$|a_k| + |a_{n-k+1}| \geqslant |a_{n-k+1} - a_k| \geqslant |n+1-2k|$$

for $1 \leqslant k \leqslant n$. So

$$\begin{aligned}\sum_{k=1}^{n} |a_k|^3 &= \frac{1}{2}\sum_{k=1}^{n}(|a_k|^3 + |a_{n+1-k}|^3)\\ &= \frac{1}{2}\sum_{k=1}^{n}(|a_k| + |a_{n+1-k}|)\Big(\frac{3}{4}(|a_k| - |a_{n+1-k}|)^2\\ &\quad + \frac{1}{4}(|a_k| + |a_{n+1-k}|)^2\Big)\\ &\geqslant \frac{1}{8}\sum_{k=1}^{n}(|a_k| + |a_{n+1-k}|)^3\\ &\geqslant \frac{1}{8}\sum_{k=1}^{n}|n+1-2k|^3.\end{aligned}$$

When n is odd,

$$\sum_{k=1}^{n}|n+1-2k|^3 = 2\times 2^3 \times \sum_{i=1}^{\frac{n-1}{2}} i^3 = \frac{1}{4}(n^2-1)^2.$$

When n is even,

$$\begin{aligned}\sum_{k=1}^{n} |n+1-2k|^3 &= 2\sum_{i=1}^{\frac{n}{2}} (2i-1)^3 \\ &= 2\left(\sum_{j=1}^{n} j^3 - \sum_{i=1}^{\frac{n}{2}} (2i)^3\right) \\ &= \frac{1}{4}n^2(n^2-2).\end{aligned}$$

So

$$\sum_{k=1}^{n} |a_k|^3 \geqslant \frac{1}{32}(n^2-1)^2$$

for odd n, and

$$\sum_{k=1}^{n} |a_k|^3 \geqslant \frac{1}{32}n^2(n^2-2)$$

for even n. The equality holds at $a_i = i - \frac{n+1}{2}$, $i = 1, 2, \ldots, n$.

5 Find all integers n such that we can color all the edges and the diagonals of a convex n-polygon by n given colors satisfying the following conditions:

(1) Each of the edges or the diagonals is colored by only one color;

(2) For any three distinct colors, there exists a triangle whose vertices are vertices of the n-polygon and three edges are colored by these three colors. (Posed by Su Chun)

Solution Answer: Any odd number $n > 1$.

First of all, there are $\binom{n}{3}$ ways to choose three among n

colors, and $\binom{n}{3}$ ways to choose three vertices to form a triangle, so if the question's condition is fulfilled, all the triangles should have a different color combination (a $1-1$ correspondence). Note that any two segments of the same color cannot have a common endpoint.

As each color combination is used in exactly one triangle, for each color there should be exactly $\binom{n-1}{2}$ triangles which have one side in this color, and so there should be exactly $\frac{n-1}{2}$ lines of this color. Therefore, n is odd.

Now we give a construction method for any odd n.

As the orientation of the vertices does not really matter, we may assume that the polygon is a regular n-polygon. First, color the n sides of the polygon in the n distinct colors. Then, for each side, color those diagonals that are parallel to this side into the same color.

In this way, for each color, there are n diagonals colored in this color, and notice that each of these diagonals is of a different length. ①

Besides, for any two triangles with all vertices in, we shall prove that they should have different color combinations. Suppose that, on the contrary, they have exactly the same three colors as their sides. Because of that all the sides with the same line are parallel, and the two triangles must be similar. For their vertices to be in the same circle, they must be the same, but this is a contradiction of ①. This completes the proof.

6 Given an integer $n \geqslant 3$, prove that there exists a set S of n

distinct positive integers such that for any two distinct nonempty subsets A and B of S, the numbers

$$\frac{\sum_{x \in A} x}{|A|}, \frac{\sum_{x \in B} x}{|B|}$$

are two coprime composite integers, and $\sum_{x \in X} x$ denotes the sum of all elements of a finite set X, and $|X|$ denotes the cardinality of X. (Posed by Yu Hongbing)

Proof Let $f(X)$ be the average of elements of the finite number set X. First of all, make n different primes $p_1, p_2, \ldots, p_n$ which are all bigger than n, and we prove that for any different nonempty subset A, B of the set $S_1 = \left\{ \frac{\prod_{i=1}^{n} p_i}{p_j} : 1 \leqslant j \leqslant n \right\}$, $f(A) \neq f(B)$ always holds.

In fact, we can suppose that $\frac{\prod_{i=1}^{n} p_i}{p_1} \in A$ and $\frac{\prod_{i=1}^{n} p_i}{p_1} \notin B$ without loss of generality. Every element of B can be divided by p_1, so $p_1 \mid n!\, f(B)$. But A has exactly one element which cannot be divided by p_1, so we find that $n!\, f(A)$ cannot divided by p_1 (note that $p_1 > n$), and therefore $n! f(A) \neq n! f(B)$, which follows $f(A) \neq f(B)$.

Second, let $S_2 = \{n! x : x \in S_1\}$. Then $f(A)$ and $f(B)$ are different positive integers when A, B are different nonempty subsets of S_2.

In fact, it is easy to see that there exist two sets A_1, B_1 which are different nonempty subsets of S_1, and $f(A) = n!\, f(A_1)$, $f(B) = n! f(B_1)$ holds. We get $f(A) \neq f(B)$ from $f(A_1) \neq f(B_1)$, and $f(A)$, $f(B)$ are positive integers from

$|A|, |B| \leqslant n$ and their elements are all positive.

Then, let K be the largest element of S_2. We prove that for every two distinct subsets A, B of the set $S_3 = \{K!x + 1: x \in S_2\}$, $f(A)$ and $f(B)$ are coprime integers which are both larger than 1.

In fact, it is easy to see that there exist two sets A_1, B_1 which are different nonempty subsets of S_2, and $f(A) = K!f(A_1) + 1$, $f(B) = K!f(B_1) + 1$ holds. Obviously, $f(A)$ and $f(B)$ are different integers which are both larger than 1. If they have common divisors, let p be a prime common divisor of them without loss of generality. Clearly, we have $p \mid (K! \cdot |f(A_1) - f(B_1)|)$. We get $1 \leqslant |f(A_1) - f(B_1)| \leqslant K$ by $0 < f(A_1)$, $f(B_1) \leqslant K$ and $f(A_1) \neq f(B_1)$, so $p \leqslant K$, which follows $p \mid K!f(A_1)$, and then $p \mid 1$ — a contradiction.

Lastly, let L be the largest element of S_3. We prove that for every two distinct nonempty subsets A, B of the set $S_4 = \{L! + x : x \in S_3\}$, $f(A)$ and $f(B)$ are two composites which share no common divisors.

In fact, it is easy to see that there exist two sets A_1, B_1 which are different nonempty subsets of S_3, and $f(A) = L! + f(A_1)$, $f(B) = L! + f(B_1)$ holds. Obviously, $f(A)$ and $f(B)$ are different integers which are both larger than 1. Because of that, L is the largest element of S_3, and we have $f(A_1) \mid L!$ and $f(A_1) \mid f(A)$. We find that $f(A)$ is composite by $f(A_1) < f(A)$. For a similar reason, $f(B)$ is composite too. If they have common divisors, let p be a prime common divisor of them without loss of generality. It is obvious that $p \mid (L! \cdot |f(A_1) - f(B_1)|)$. We get $1 \leqslant |f(A_1) - f(B_1)| \leqslant L$ by $0 < f(A_1)$, $f(B_1) \leqslant L$ and $f(A_1) \neq f(B_1)$, so $p \leqslant L$, which follows $p \mid f(A_1)$ and $p \mid f(B_1)$ — a contradiction of the fact that $f(A_1)$ and $f(B_1)$ are coprime. This completes the proof.

2010 (Chongqing)

First Day

(0800 - 1230; January 22, 2010)

1 As shown below, two circles Γ_1, Γ_2 intersect at points A, B, one line passing through B intersects Γ_1, Γ_2 at points C, D, another line passing through B intersects Γ_1, Γ_2 at points E, F, and line CF intersects Γ_1, Γ_2 at points P, Q, respectively. Let M, N be the middle points of arc PB and arc QB, respectively. Prove that if $CD = EF$, then C, F, M, N are concyclic. (Posed by Xiong Bin)

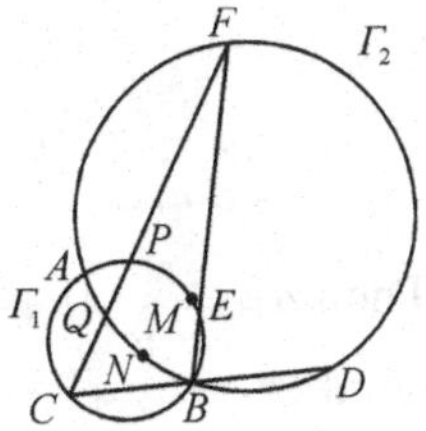

Proof Draw lines AC, AD, AE, AF, DF. From $\angle ADB = \angle AFB$, $\angle ACB = \angle AEF$ and the assumption $CD = EF$; one obtains $\triangle ACD \cong \triangle AEF$. So we have $AD = AF$, $\angle ADC = \angle AFE$ and $\angle ADF = \angle AFD$. Then $\angle ABC = \angle AFD = \angle ADF = \angle ABF$; AB is the bisector of $\angle CBF$. Draw lines CM, FN. Since M is the middle point of arc PB, CM is the bisector of $\angle DCF$, and FN is the bisector of $\angle CFB$. Then BA, CM, FN have a common intersection, say, at I. In the circles Γ_1 and Γ_2, one has $CI \times IM = AI \times IB$, $AI \times IB = NI \times IF$, according to the theorem of power with respect to circles. So $NI \times IF = CI \times IM$, and C, F, M, N are concyclic.

2 Given an integer $k \geqslant 3$ and a sequence $\{a_n\}$ that satisfies $a_k = 2k$ and for each $n > k$,

$$a_n = \begin{cases} a_{n-1} + 1, & \text{if } a_{n-1} \text{ and } n \text{ are coprime,} \\ 2n, & \text{otherwise.} \end{cases}$$

Prove that $a_n - a_{n-1}$ is a prime for infinitely many n. (Posed by Zhu Huawei)

Proof Suppose that $a_l = 2l$, $l \geqslant k$. Let p be the least prime divisor of $k - 1$. Then $(l - 1, i) = \begin{cases} 1, & 1 \leqslant i < p, \\ p, & i = p, \end{cases}$ and thus

$$(2l + i - 2, l + i - 1) = \begin{cases} 1, & 1 \leqslant i < p, \\ p, & i = p. \end{cases}$$

From (1) we know that

$$a_{l+i-1} = \begin{cases} 2l + i - 1, & 1 \leqslant i < p, \\ 2l + 2p - 2, & i = p. \end{cases}$$

Then $a_{l+p-1} - a_{l+p-2} = (2l + 2p - 2) - (2l + p - 2) = p$ is a prime number, $a_{l+p-1} = 2(l + p - 1)$. From the discussion above, we know that there are infinitely many $l \geqslant k$, such that $a_l = 2l$ and $a_{l+p-1} - a_{l+p-2} = p$ is the least prime divisor of $l - 1$.

3 Let a, b, c be complex numbers such that $|az^2 + bz + c| \leqslant 1$ for all complex numbers z with $|z| \leqslant 1$. Find the maximum of $|bc|$. (Posed by Li Weigu)

Solution Write $f(z) = az^2 + bz + c$. We first prove that

$$|f(z)| \leqslant 1 \text{ for all } z, |z| \leqslant 1 \Leftrightarrow |f(z)| \leqslant 1 \text{ for all } z, |z| = 1. \quad ①$$

Assume that $f(z) = a(z - \alpha)(z - \beta)$. For any z, $|z| < 1$, if $\alpha = \beta$, one of the two intersection points of the line through α

and the origin with the unit circle is closer to α than z is; if $\alpha \neq \beta$, the line through z and perpendicular to the line through α, β intersects the unit circles at two points, one of which is closer to α, β than z is, respectively. So ① holds.

For any complex number z, $|z| = 1$, it is obvious that

$$|f(z)| = |cz^{-2} + bz^{-1} + a|.$$

So $|ab|_{\max} = |bc|_{\max}$. Write $a'z^2 + b'z + c' = e^{i\alpha} f(e^{i\beta} z)$. One can choose real numbers α, β such that a', b' are positive real numbers, so one can assume that a, $b \geqslant 0$ without loss of generality

$$1 \geqslant |f(e^{i\theta})| \geqslant |\operatorname{Im} f(e^{i\theta})| = |a\sin 2\theta + b\sin\theta + \operatorname{Im} c|.$$

Without loss of generality we can assume $\operatorname{Im} c \geqslant 0$ (otherwise take a map, $\theta \to -\theta$). For any $\theta \in \left(0, \frac{\pi}{2}\right)$,

$$1 \geqslant a\sin 2\theta + b\sin\theta \geqslant 2\sqrt{ab\sin 2\theta\sin\theta}$$

$$\Rightarrow ab \leqslant \frac{1}{4\sin 2\theta\sin\theta}, \ \theta \in \left(0, \frac{\pi}{2}\right)$$

$$\Rightarrow ab \leqslant \min_{\theta\in\left(0,\frac{\pi}{2}\right)} \frac{1}{4\sin 2\theta\sin\theta} = \frac{1}{4\max\limits_{\theta\in\left(0,\frac{\pi}{2}\right)}(\sin 2\theta\sin\theta)} = \frac{3\sqrt{3}}{16}$$

$$\Rightarrow |bc|_{\max} = |ab|_{\max} \leqslant \frac{3\sqrt{3}}{16}.$$

An example of $|bc| = \frac{3\sqrt{3}}{16}$ is

$$f(z) = \frac{\sqrt{2}}{8}z^2 - \frac{\sqrt{6}}{4}z - \frac{3\sqrt{2}}{8},$$

$$|f(e^{i\theta})|^2 = 1 - \frac{3}{8}\left(\cos\theta - \frac{\sqrt{3}}{3}\right)^2 \leqslant 1.$$

Second Day

(0800 - 1230; January 23, 2010)

4 Given two integers m, n greater than 1, and integers $a_1 < a_2 < \cdots < a_m$, prove that there exists a set T of integers with $|T| \leqslant 1 + \frac{a_m - a_1}{2n+1}$ such that each a_i can be written as $a_i = t + s$ for some $t \in T$, and $s \in [-n, n]$. (Posed by Leng Gangsong)

Proof Write $a_1 = a$, $a_m = b$, $b - a = (2n+1)q + r$, where q, $r \in \mathbf{Z}$ and $0 \leqslant r \leqslant 2n$. Take

$$T = \{a + n + (2n+1)k \mid k = 0, 1, \ldots, q\}.$$

Then $|T| = q + 1 \leqslant 1 + \frac{b-a}{2n+1}$. We have the set

$$\begin{aligned} B &= \{t + s \mid t \in T, s = -n, -n+1, \ldots, n\} \\ &= \{a, a+1, \cdots, a + (2n+1)q + 2n\}. \end{aligned}$$

Note that $a + (2n+1)q + 2n \geqslant a + (2n+1)q + r = b$, so each a_i belongs to B.

5 We operate on piles of cards placed at $n+1$ positions A_1, A_2, ..., A_n $(n \geqslant 3)$ and O. In one operation, we can do either of the following:

(1) If there are at least three cards at A_i, we may take three cards from A_i and place one at each of A_{i-1}, A_{i+1} and O (assume that $A_0 = A_n$, $A_{n+1} = A_1$);

(2) If there are at least n cards at O, we may take n cards from O and place one at each of A_1, A_2, ..., A_n.

Prove that if the total number of cards is at least $n^2 + 3n + 1$, we can take some operations such that there are at least

$n+1$ cards at each position. (Posed by Qu Zhenhua)

Proof One only needs to consider the case with the total number of cards equal n^2+3n+1. We take the following strategy. If there are at least three cards at some A_i, then use operation (1) at this position. Such operations can be done for only finitely many steps. Then we have no more than two cards at each A_i and no less than n^2+n+1 at O.

We then take operation (2) for $n+1$ times. There are then at least $n+1$ cards at each A_i. We will now prove that one can increase the number of cards at O to at least $n+1$, while keeping at least $n+1$ cards at each A_i.

Put $A_1, A_2, \ldots, A_n$ evenly and in increasing order on a circle with the center at O. We call a group of consecutive A_i's, $G=\{A_i, A_{i+1}, \ldots, A_{i+l-1}\}$, $1 \leqslant i \leqslant n$, $1 \leqslant l \leqslant n$, on the circle a team, where for $j>n$ we define $A_j=A_{j-n}$. A team is good if after we take operation (1) once at each point in G, there are at least $n+1$ cards at every point in G. Write $a_1, a_2, \ldots, a_n$ as the number of cards at points $A_1, A_2, \ldots, A_n$, $a_i \geqslant n+1$, $i=1, 2, \ldots, n$. Let $G=\{A_i\}$ be a team. Then, if there is only one point A_i in G, G is good iff $a_i \geqslant n+4$; a team of two points $G=\{A_i, A_{i+1}\}$ is good iff a_i, $a_{i+1} \geqslant n+3$; a team $G=\{A_i, A_{i+1}, \ldots, A_{i+l-1}\}$ of l points $(3 \leqslant l \leqslant n-1)$ is good iff a_i, $a_{i+l-1} \geqslant n+3$ and $a_j \geqslant n+2$, $i+1 \leqslant j \leqslant i+l-2$; finally, the team $G=\{A_1, A_2, \ldots, A_n\}$ of all n points is good iff $a_j \geqslant n+2$, $1 \leqslant j \leqslant n$. We then prove that there must exist at least one good team if $a_1+a_2+\cdots+a_n \geqslant n^2+2n+1$.

Assume that there is no good team. Then each $a_i \in \{n+1, n+2, n+3\}$, otherwise there is a good team of one point at any A_i with $a_i \geqslant n+4$. Suppose that the number of $n+1$, $n+2$ and $n+3$'s among $a_1, a_2, \ldots, a_n$ are x, y, z respectively. We will show that $x \geqslant z$. Since $n^2 + 2n + 1 > n(n+2)$, $z \geqslant 1$. If $z = 1$, then $x \geqslant 1$, otherwise all $a_i \geqslant n+2$ and $G = \{A_1, A_2, \ldots, A_n\}$ is a good team. If $z \geqslant 2$, the z points with $n+3$ cards divide the circle into z arcs (no two of these z points are adjacent). Since there is no good team by assumption, there is at least one point on each arc with $n+1$ cards. So $x \geqslant z$, and the total number of cards at $A_1, A_2, \ldots, A_n$ is

$$\begin{aligned} x(n+1) + y(n+2) + z(n+3) &\leqslant (x+y+z)(n+2) \\ &= n(n+2) < n^2 + 2n + 1, \end{aligned}$$

which is a contradiction. Thus, we have proven that when the number of cards at O is less than $n+1$, there exists a good team. We use operation (1) on each point of a good team; the number of cards at O is increased while the number of cards at each $A_1, A_2, \ldots, A_n$ is still no less than $n+1$. We can reiterate until there are at least $n+1$ cards at O.

6 Let $a_1, a_2, a_3, b_1, b_2, b_3$ be pairwise distinct positive integers such that

$$(n+1)a_1^n + na_2^n + (n-1)a_3^n \mid (n+1)b_1^n + nb_2^n + (n-1)b_3^n$$

holds for all positive integers n. Prove that there exists a positive integer k such that $b_i = ka_i$ for all $i = 1, 2, 3$. (Posed by Chen Yonggao)

Proof Suppose that r is any positive integer. Since there are infinitely many primes, there is a prime p, such that

$$p > (a_1^r + a_2^r + a_3^r)(b_1^r + b_2^r + b_3^r). \qquad ①$$

Because p is prime and ①, we have $(p, a_1^r + a_2^r + a_3^r) = 1$. p is coprime to $p-1$; from the Chinese remainder theorem, there is a positive integer n such that

$$n \equiv r \pmod{p-1}, \quad ②$$

$$n(a_1^r + a_2^r + a_3^r) + a_1^r - a_3^r \equiv 0 \pmod{p}. \quad ③$$

From ②, ③ and Fermat's theorem,

$$(n+1)a_1^n + na_2^n + (n-1)a_3^n \equiv n(a_1^r + a_2^r + a_3^r) + a_1^r - a_3^r$$
$$\equiv 0 \pmod{p}. \quad ④$$

From the assumption of the problem,

$$(n+1)b_1^n + nb_2^n + (n-1)b_3^n \equiv 0 \pmod{p}.$$

Again from ② and Fermat's little theorem,

$$n(b_1^r + b_2^r + b_3^r) + b_1^r - b_3^r \equiv 0 \pmod{p}. \quad ⑤$$

Eliminate n from ④, ⑤:

$$(a_1^r + a_2^r + a_3^r)(b_1^r - b_3^r) \equiv (b_1^r + b_2^r + b_3^r)(a_1^r - a_3^r) \pmod{p}. \quad ⑥$$

From ①, ⑥ we have

$$(a_1^r + a_2^r + a_3^r)(b_1^r - b_3^r) = (b_1^r + b_2^r + b_3^r)(a_1^r - a_3^r),$$

and thus

$$(a_2b_1)^r + 2(a_3b_1)^r + (a_3b_2)^r = (a_1b_2)^r + 2(a_1b_3)^r + (a_2b_3)^r. \quad ⑦$$

We then prove the following lemma.

Lemma *Assume that* $x_1, \ldots, x_s, y_1, \ldots, y_s$ *are real numbers*,

$$0 < x_1 \leqslant x_2 \leqslant \cdots \leqslant x_s,\ 0 < y_1 \leqslant y_2 \leqslant \cdots \leqslant y_s,$$

such that for any positive integers r,

$$x_1^r + x_2^r + \cdots + x_s^r = y_1^r + y_2^r + \cdots + y_s^r.$$

Then $x_i = y_i (i = 1, 2, \ldots, s)$.

Proof of the lemma. We use induction on s. If $s = 1$, take $r = 1$; then $x_1 = y_1$. Assume that the lemma holds when $s = t$.

When $s = t + 1$, if $x_{t+1} \neq y_{t+1}$, say $x_{t+1} < y_{t+1}$,

$$\left(\frac{x_1}{y_{t+1}}\right)^r + \cdots + \left(\frac{x_{t+1}}{y_{t+1}}\right)^r = \left(\frac{y_1}{y_{t+1}}\right)^r + \cdots + \left(\frac{y_t}{y_{t+1}}\right)^r + 1 \geqslant 1.$$

Because $0 < \frac{x_i}{y_{t+1}} < 1$ $(1 \leqslant i \leqslant t + 1)$, take the limit $r \to +\infty$, and we have $0 \geqslant 1$, a contradiction.

So $x_{t+1} = y_{t+1}$, and then $x_1^r + \cdots + x_t^r = y_1^r + \cdots + y_t^r$, $r = 1, 2, \ldots$. By induction the lemma holds for all positive integer s.

Now return to the proof of the problem. Since a_1, a_2, a_3, b_1, b_2, b_3 are distinct,

$$a_2b_1 \neq a_3b_1,\ a_3b_1 \neq a_3b_2,\ a_1b_2 \neq a_1b_3,$$
$$a_1b_3 \neq a_2b_3,\ a_2b_1 \neq a_2b_3.$$

From ⑦ and the lemma we know that

$$a_2b_1 = a_3b_2 = a_1b_3,\ a_3b_1 = a_1b_2 = a_2b_3, \tag{8}$$

or

$$a_2b_1 = a_1b_2,\ a_3b_1 = a_1b_3,\ a_3b_2 = a_2b_3. \tag{9}$$

If ⑧ holds, then $\frac{a_2b_1}{a_3b_1} = \frac{a_3b_2}{a_1b_2} = \frac{a_1b_3}{a_2b_3}$, i.e. $\frac{a_2}{a_3} = \frac{a_3}{a_1} = \frac{a_1}{a_2}$, which is a contradiction to a_1, a_2, a_3 being distinct. So ⑨ holds. Then $\frac{b_1}{a_1} = \frac{b_2}{a_2} = \frac{b_3}{a_3}$. Write $\frac{b_1}{a_1} = \frac{k}{l}$, $(k, l) = 1$, $l \geqslant 1$; then $b_i = \frac{k}{l}a_i$, $i = 1, 2, 3$. From $2b_1 + b_2 = \frac{k}{l}(2a_1 + a_2)$ and the assumption of the problem (with $n = 1$), $2a_1 + a_2 \mid 2b_1 + b_2$ and $\frac{k}{l}$ is an integer. So $l = 1$ and $b_i = ka_i$, $i = 1, 2, 3$.

China National Team Selection Test

2009 (Wuhan, Hubei)

First Day

(0800 - 1230; March 31, 2009)

1 Let D be a point on side BC of triangle ABC such that $\angle CAD = \angle CBA$. A circle with center O passes through B, D, and meets segments AB, AD at E, F, respectively. Lines BF and DE meet at point G. M is the midpoint of AG. Prove that $CM \perp AO$. (Posed by Xiong Bin)

Proof As shown in Fig. 1, extend EF and meet BC at point

P, and join and extend GP, which meets AD at K and the extension of AC at L.

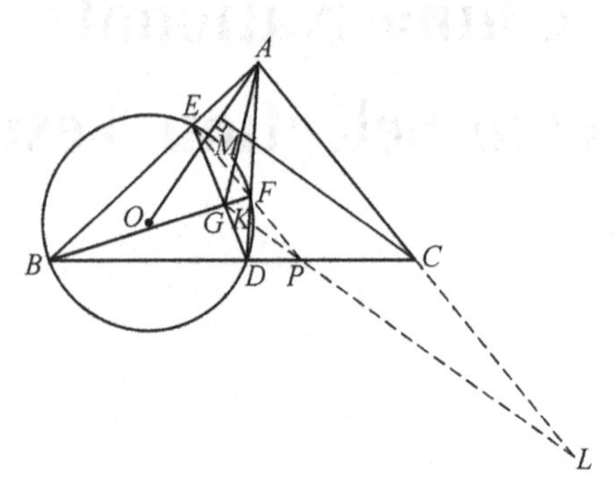

Fig. 1

As shown in Fig. 2, let Q be a point on AP such that

$\angle PQF = \angle AEF = \angle ADB$.

It is easy to see that A, E, F, Q and F, D, P, Q are concyclic respectively. Denote by r the radius of $\odot O$. By the power of a point theorem,

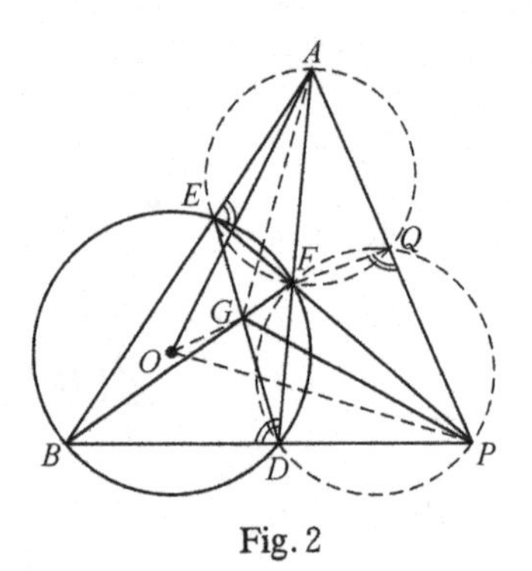

Fig. 2

$$\begin{aligned} AP^2 &= AQ \times AP + PQ \times AP \\ &= AF \times AD + PF \times PE \\ &= (AO^2 - r^2) + (PO^2 - r^2). \end{aligned} \quad ①$$

Similarly,

$$AG^2 = (AO^2 - r^2) + (GO^2 - r^2). \quad ②$$

By ①, ②, we have $AP^2 - AG^2 = PO^2 - GO^2$, which implies that $PG \perp AO$. As shown in Fig. 3, applying Menelaus' theorem for $\triangle PFD$ and line AEB, we have

$$\frac{DA}{AF} \times \frac{FE}{EP} \times \frac{PB}{BD} = 1. \quad ③$$

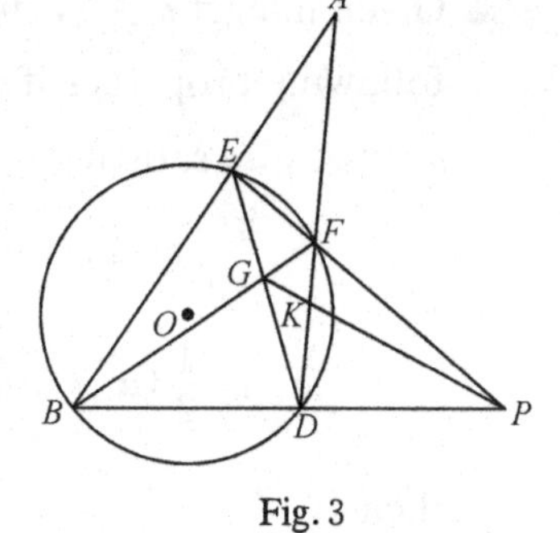

Fig. 3

Applying Ceva's theorem for $\triangle PFD$ and point G, we have

$$\frac{DK}{KF} \times \frac{FE}{EP} \times \frac{PB}{BD} = 1. \quad ④$$

③ ÷ ④ yields

$$\frac{DA}{AF} = \frac{DK}{KF}. \quad ⑤$$

Equation ⑤ illustrates that A, K are harmonic points with respect to F, D, i.e. $AF \times KD = AD \times FK$.

It follows that

$$AK \times FD = AF \times KD + AD \times FK = 2AF \times KD. \quad ⑥$$

Since B, D, F, E are concyclic, we have $\angle DBA = \angle EFA$. Since $\angle CAD = \angle CBA$, we have $\angle CAF = \angle EFA$, which implies that $AC \parallel EP$. Thus,

$$\frac{CP}{PD} = \frac{AF}{FD}. \quad ⑦$$

Applying Menelaus' theorem to $\triangle ACD$ and line LPK, we have

$$\frac{AL}{LC} \times \frac{CP}{PD} \times \frac{DK}{KA} = 1. \quad ⑧$$

Combining ⑥, ⑦ and ⑧, we have $\frac{AL}{LC} = 2$.

In $\triangle AGL$, M, C are the midpoints of AG, AL respectively, and hence $MC \parallel GL$. Since $GL \perp AO$, we conclude that $MC \perp AO$.

2 Given integer $n \geqslant 2$, find the largest number $\lambda(n)$ with the following property: if a sequence of real numbers $a_0, a_1, a_2, \ldots, a_n$ satisfies

$$0 = a_0 \leqslant a_1 \leqslant a_2 \leqslant \cdots \leqslant a_n,$$

$$a_i \geqslant \frac{1}{2}(a_{i+1} + a_{i-1}),\ i = 1, 2, \ldots, n-1,$$

then

$$\left(\sum_{i=1}^{n} ia_i\right)^2 \geqslant \lambda(n) \sum_{i=1}^{n} a_i^2.$$

(Posed by Zhu Huawei)

Solution The largest possible value of $\lambda(n)$ is $\frac{n(n+1)^2}{4}$.

Let $a_1 = a_2 = \cdots = a_n = 1$. Then we have $\lambda(n) \leqslant \frac{n(n+1)^2}{4}$.

We shall show that for any real numbers $a_0, a_1, a_2, \ldots, a_n$ satisfying the indicated property in the problem, the following inequality holds:

$$\left(\sum_{i=1}^{n} ia_i\right)^2 \geqslant \frac{n(n+1)^2}{4}\left(\sum_{i=1}^{n} a_i^2\right). \qquad ①$$

First, we notice that

$$a_1 \geqslant \frac{a_2}{2} \geqslant \cdots \geqslant \frac{a_n}{n}.$$

Indeed, by assumption, $2ia_i \geqslant i(a_{i+1} + a_{i-1})$ holds for $i = 1, 2, \ldots, n-1$. For any given positive integer $1 \leqslant l \leqslant n-1$, summing the above inequality over $i = 1, 2, \ldots, l$, we have $(l+1)a_l \geqslant la_{l+1}$, i.e.

$$\frac{a_l}{l} \geqslant \frac{a_{l+1}}{l+1} \text{ for } l = 1, 2, \ldots, n-1.$$

In what follows, we show that for any i, j, $k \in \{1, 2, \ldots, n\}$, if $i > j$, then

$$\frac{2ik^2}{i+k} > \frac{2jk^2}{j+k}.$$

Indeed, the above inequality is equivalent to $2ik^2(j+k) > 2jk^2(i+k)$, i.e. $(i-j)k^3 > 0$, which is clearly true.

Now, we are going to show inequality ①. We shall start by estimating the lower bound of $a_i a_j$ for $1 \leqslant i < j \leqslant n$.

By previous results, we have $\frac{a_i}{i} \geqslant \frac{a_j}{j}$, i.e. $ja_i - ia_j \geqslant 0$. Since $a_i - a_j \leqslant 0$, we have $(ja_i - ia_j)(a_j - a_i) \geqslant 0$, i.e. $a_i a_j \geqslant \frac{i}{i+j} a_j^2 + \frac{j}{i+j} a_i^2$.

Thus, we have

$$\begin{aligned}\left(\sum_{i=1}^{n} ia_i\right)^2 &= \sum_{i=1}^{n} i^2 a_i^2 + 2 \sum_{1\leqslant i<j\leqslant n} ija_i a_j \\ &\geqslant \sum_{i=1}^{n} i^2 \times a_i^2 + 2 \sum_{1\leqslant i<j\leqslant n} \left(\frac{i^2 j}{i+j} a_j^2 + \frac{ij^2}{i+j} a_i^2\right) \\ &= \sum_{i=1}^{n} \left(a_i^2 \times \sum_{k=1}^{n} \frac{2ik^2}{i+k}\right).\end{aligned}$$

Let $b_i = \sum_{k=1}^{n} \frac{2ik^2}{i+k}$. We see from previous results that $b_1 \leqslant b_2 \leqslant \cdots \leqslant b_n$.

Since $a_1^2 \leqslant a_2^2 \leqslant \cdots \leqslant a_n^2$, by the Chebyshev inequality, we have

$$\sum_{i=1}^{n} a_i^2 b_i \geqslant \frac{1}{n}\left(\sum_{i=1}^{n} a_i^2\right)\left(\sum_{i=1}^{n} b_i\right).$$

Hence $\left(\sum_{i=1}^{n} ia_i\right)^2 \geqslant \frac{1}{n}\left(\sum_{i=1}^{n} a_i^2\right)\left(\sum_{i=1}^{n} b_i\right)$.

Since

$$\sum_{i=1}^{n} b_i = \sum_{i=1}^{n}\sum_{k=1}^{n} \frac{2ik^2}{i+k} = \sum_{i=1}^{n} i^2 + 2\sum_{1\leqslant i<j\leqslant n}\left(\frac{i^2 j}{i+j} + \frac{ij^2}{i+j}\right)$$

$$= \sum_{i=1}^{n} i^2 + 2\sum_{1\leqslant i<j\leqslant n} ij = \left(\sum_{i=1}^{n} i\right)^2 = \frac{n^2(n+1)^2}{4},$$

we find that $\left(\sum_{i=1}^{n} ia_i\right)^2 \geqslant \frac{n(n+1)^2}{4}\sum_{i=1}^{n} a_i^2$, which proves inequality ①.

We conclude that the maximum possible value of $\lambda(n)$ is $\frac{n(n+1)^2}{4}$.

3 Prove that for any odd prime number p, the number of positive integers n satisfying $p \mid n! + 1$ is no more than $cp^{\frac{2}{3}}$, where c is a constant number independent of p. (Posed by Yu Hongbing)

Proof Clearly, if n satisfies the required property, then $1 \leqslant n \leqslant p-1$. Denote all such n's by $n_1 < n_2 < \cdots < n_k$; we shall show that $k \leqslant 12p^{\frac{2}{3}}$. If $k \leqslant 12$ there is nothing to prove. In what follows, we assume that $k > 12$.

Rename $n_{i+1} - n_i (1 \leqslant i \leqslant k-1)$ in nondecreasing order as $1 \leqslant \mu_1 \leqslant \mu_2 \leqslant \cdots \leqslant \mu_{k-1}$. It is clear that

$$\sum_{i=1}^{k-1} \mu_i = \sum_{i=1}^{k} (n_{i+1} - n_i) = n_k - n_1 < p. \qquad ①$$

First, we show that for any $s \geqslant 1$,

$$|\{1 \leqslant i \leqslant k-1 : \mu_i = s\}| \leqslant s, \qquad ②$$

i.e. there are at most s μ_i's equal to s.

In fact, suppose that $n_{i+1} - n_i = s$; then $n_i! + 1 \equiv n_{i+1}! + 1 \equiv 0 \pmod{p}$, so $(p, n_i!) = 1$, and

$$(n_i + s)(n_i + s - 1)\cdots(n_i + 1) \equiv 1 \pmod p.$$

Thus, n_i is a solution to the congruence equation

$$(x + s)(x + s - 1)\cdots(x + 1) \equiv 1 \pmod p.$$

Since p is a prime number, there are at most s solutions to the above equation, thanks to Lagrange's theorem. Thus, there are at most s n_i's with $n_{i+1} - n_i = s$, i.e. ② holds.

Now, we show that for any nonnegative integer l, if $\frac{l(l+1)}{2} + 1 \leqslant k - 1$, then $\mu_{\frac{l(l+1)}{2}+1} \geqslant l + 1$. Suppose on the contrary that $\mu_{\frac{l(l+1)}{2}+1} \leqslant l$. Then $\mu_1, \mu_2, \ldots, \mu_{\frac{l(l+1)}{2}+1}$ are all positive integers between 1 and l. By ②, there is at most one μ_i equal to 1, at most two μ_i's equal to 2, ..., at most l μ_i's equal to l, and thus there are at most $1 + 2 + \cdots + l = \frac{l(l+1)}{2}$ μ_i's less than or equal to l, which contradicts the assumption that $\mu_1, \mu_2, \ldots, \mu_{\frac{l(l+1)}{2}+1}$ are all less than or equal to l.

Let m be the largest positive integer with $\frac{m(m+1)}{2} + 1 \leqslant k - 1$. Then

$$\frac{m(m+1)}{2} + 1 \leqslant k - 1 < \frac{(m+1)(m+2)}{2} + 1, \qquad ③$$

and hence

$$\begin{aligned}\sum_{i=1}^{k-1} \mu_i &\geqslant \sum_{i=0}^{m-1} \left(\mu_{\frac{i(i+1)}{2}+1} + \mu_{\frac{i(i+1)}{2}+2} + \cdots + \mu_{\frac{(i+1)(i+2)}{2}}\right) \\ &\geqslant \sum_{i=0}^{m-1} (i+1)\mu_{\frac{i(i+1)}{2}+1} \geqslant \sum_{i=0}^{m-1} (i+1)^2 \\ &= \frac{m(m+1)(2m+1)}{6} > \frac{m^3}{3}.\end{aligned}$$

Since $k > 12$, $m \geqslant 4$, combining ① and ③ we get

$$k < 2 + \frac{(m+1)(m+2)}{2}$$

$$< 4m^2 < 4\left(3\sum_{i=1}^{k-1}\mu_i\right)^{\frac{2}{3}}$$

$$< 4 \times (3p)^{\frac{2}{3}}.$$

This completes our proof.

Second Day

(0800 - 1230; April 1, 2009)

4 Let a, b be positive real numbers with $b - a > 2$. Prove that for any two distinct integers m, n in the interval $[a, b)$, there is a nonempty set S consisting of some integers in the interval $[ab, (a+1)(b+1))$, such that $\frac{\prod_{x \in S} x}{mn}$ is a square of a rational number. (Posed by Yu Hongbing)

Proof We first prove the following lemma:

Lemma *Let u be an integer with $a \leqslant u < u+1 < b$. Then there are two distinct integers x, y in the interval $[ab, (a+1) \times (b+1))$, such that $\frac{xy}{u(u+1)}$ is a square of an integer.*

Proof of lemma. Let v be the smallest integer not less than $\frac{ab}{u}$, i.e. v satisfies

$$\frac{ab}{u} \leqslant v < \frac{ab}{u} + 1;$$

hence

$$ab \leqslant uv < ab + u (< ab + a + b + 1), \qquad ①$$

and thus

$$ab < (u+1)v = uv + v < ab + u + \frac{ab}{u} + 1 < ab + a + b + 1 (\text{since } a \leqslant u < b). \quad ②$$

(Here we have used a well-known result: the function $f(t) = t + \frac{ab}{t}$ $(a \leqslant t \leqslant b)$ attains its maximum at $t = a$ or b.)

By ① and ②, we see that uv and $(u+1)v$ are two distinct integers in the interval $I = [ab, (a+1)(b+1))$. Let $x = uv$ and $y = (u+1)v$. Then $\frac{xy}{u(u+1)} = v^2$ is a square of an integer number. We have verified the lemma.

Going back to the original problem, suppose that $m < n$. Then $a \leqslant m \leqslant n-1 < b$. It follows from the lemma that for every $k = m, m+1, \ldots, n-1$, there exist x_k, y_k, two distinct integers in the interval $[ab, (a+1)(b+1))$, and integer A_k, such that

$$\frac{x_k y_k}{k(k+1)} = A_k^2.$$

Multiplying all together, we find that

$$\frac{\prod_{k=m}^{n-1} x_k y_k}{mn(m+1)^2 \cdots (n-1)^2} = \prod_{k=m}^{n-1} A_k^2$$

is a square of an integer.

Let S be the set of numbers that appear in x_i, y_i $(m \leqslant i \leqslant n-1)$ odd times. If S is nonempty, then it follows from the above equality that $\frac{\prod_{x \in S} x}{mn}$ is a square of a rational number.

If S is empty, then mn is a square of an integer. Since $a + b > 2\sqrt{ab}$, we have $ab + a + b + 1 > ab + 2\sqrt{ab} + 1$, i.e. $\sqrt{(a+1)(b+1)} > \sqrt{ab} + 1$, which means that there is at least an integer in the interval $[\sqrt{ab}, \sqrt{(a+1)(b+1)})$. As a result there is a perfect square in the interval $[ab, (a+1)(b+1))$. Suppose that $r^2 \in [ab, (a+1)(b+1))$ $(r \in \mathbf{Z})$, and let $S' = \{r^2\}$. Then $\frac{\prod_{x \in S'} x}{mn}$ is a square of a rational number.

5 Let m be an integer greater than 1, and let n be an odd number with $3 \leqslant n < 2m$. Numbers $a_{i,j}$ ($i, j \in \mathbf{N}$, $1 \leqslant i \leqslant m$, $1 \leqslant j \leqslant n$) satisfy:

(1) For every $1 \leqslant j \leqslant n$, $a_{1,j}, a_{2,j}, \ldots, a_{m,j}$ is a permutation of $1, 2, \ldots, m$;

(2) $|a_{i,j} - a_{i,j+1}| \leqslant 1$ for every $1 \leqslant i \leqslant m$, $1 \leqslant j \leqslant n-1$.

Find the minimal possible value of $M = \max_{1 \leqslant i \leqslant m} \sum_{j=1}^{n} a_{i,j}$.

(Posed by Fu Yunhao)

Solution Let $n = 2l + 1$. Since $3 \leqslant n < 2m$, we have $1 \leqslant l \leqslant m - 1$. We first estimate the lower bound of M.

By the condition (1), there exists a unique $1 \leqslant i_0 \leqslant m$, such that $a_{i_0, l+1} = m$. Consider $a_{i_0, l}$ and $a_{i_0, l+2}$.

Case 1: At least one of $a_{i_0, l}$ and $a_{i_0, l+2}$ is m, and we may assume without loss of generality that $a_{i_0, l} = m$. It follows from the condition (2) that

$$a_{i_0, l-1} \geqslant m - 1,\ a_{i_0, l-2} \geqslant m - 2,\ \ldots,\ a_{i_0, 1} \geqslant m - l + 1,$$
$$a_{i_0, l+2} \geqslant m - 1,\ a_{i_0, l+3} \geqslant m - 2,\ \ldots,\ a_{i_0, 2l+1} \geqslant m - l.$$

Thus,

$$\begin{aligned} M &\geqslant \sum_{j=1}^{n} a_{i_0, j} \\ &\geqslant (m - l) + 2((m - l + 1) + (m - l + 2) + \cdots + m) \\ &= (2l + 1)m - l^2. \end{aligned}$$

Case 2: None of $a_{i_0, l}$ and $a_{i_0, l+2}$ is m, and by the condition (1) there exists $1 \leqslant i_1 \leqslant m$, $i_1 \neq i_0$, such that $a_{i_1, l} = m$. It easily follows from the conditions (1) and (2) that $a_{i_1, l+1} = m - 1$, $a_{i_1, l+2} = m$. It follows again from the condition (2) that

$$a_{i_1, l-1} \geqslant m - 1,\ a_{i_1, l-2} \geqslant m - 2,\ \ldots,\ a_{i_1, 1} \geqslant m - l + 1,$$
$$a_{i_1, l+3} \geqslant m - 1,\ a_{i_1, l+4} \geqslant m - 2,\ \ldots,\ a_{i_1, 2l+1} \geqslant m - l + 1.$$

Thus,

$$\begin{aligned} M &\geqslant \sum_{j=1}^{n} a_{i_1, j} \geqslant 2((m - l + 1) + (m - l + 2) + \cdots + m) + (m - 1) \\ &= (2l + 1)m - (l^2 - l + 1). \end{aligned}$$

Combining the above two case, we have $M \geqslant (2l + 1)m - l^2$.

On the other hand, consider

$$a_{i,j} = f(2i + j) = \begin{cases} 2i + j, & 2i + j \leqslant m, \\ (2m + 1) - (2i + j), & m + 1 \leqslant 2i + j \leqslant 2m, \\ (2i + j) - 2m, & 2m + 1 \leqslant 2i + j \leqslant 3m, \\ (4m + 1) - (2i + j), & 3m + 1 \leqslant 2i + j \leqslant 4m. \end{cases}$$

We shall show that this table of numbers satisfies the required conditions in the problem.

For any $1 \leqslant i \leqslant m$, $1 \leqslant j \leqslant n - 1$, if $m \nmid 2i + j$, then

$$| a_{i,j} - a_{i,j+1} | = | f(2i + j) - f(2i + j + 1) | = 1;$$

if $m \mid 2i + j$, then

$$| a_{i,j} - a_{i,j+1} | = | f(2i + j) - f(2i + j + 1) | = 0.$$

The condition (2) is verified.

We next show that the condition (1) is also fulfilled. In fact, it suffices to show that for any integers $1 \leqslant j \leqslant n$ and $1 \leqslant k \leqslant m$, there exists an integer $1 \leqslant i \leqslant m$ such that $a_{i,j} = k$.

If $j \equiv k \pmod 2$, since $1 \leqslant j \leqslant n$, $1 \leqslant k \leqslant m$ and $n < 2m$, we have $-2m < k - j < m$, and therefore $-m < \frac{k-j}{2} < m$, and note that $\frac{k-j}{2}$ is an integer. Thus, at least one of $\frac{k-j}{2}$ and $\frac{k-j}{2} + m$ is in the set $\{1, 2, \ldots, m\}$; taking this number to be i will do the job.

If $j \not\equiv k \pmod 2$, again since $1 \leqslant j \leqslant n$, $1 \leqslant k \leqslant m$ and $n < 2m$, we have

$$-2m < (2m+1) - (j+k) < 2m,$$

and therefore

$$-m < \frac{(2m+1)-(j+k)}{2} < m,$$

and $\frac{(2m+1)-(j+k)}{2}$ is an integer. Thus, at least one of $\frac{(2m+1)-(j+k)}{2}$ and $\frac{(2m+1)-(j+k)}{2} + m$ is in the set $\{1, 2, \ldots, m\}$; taking this number to be i will do the job.

We now estimate M in this case. Since the condition (1) holds, for any $1 \leqslant i_1 < i_2 \leqslant m$, $1 \leqslant j \leqslant n-1$, we have $f(2i_1 + j) \neq f(2i_2 + j)$, i.e. for any integers x, y with the same parity such that $3 \leqslant x < y \leqslant 2m + n$ and $y - x < 2m$, we have $f(x) \neq f(y)$. Thus, for a given i, $a_{i,1}, a_{i,3}, \ldots, a_{i,2l+1}$ are pairwise distinct, and so are numbers $a_{i,2}, a_{i,4}, \ldots, a_{i,2l}$. As a result,

$$\begin{aligned}\sum_{j=1}^{n} a_{i,j} &\leqslant (m-l) + 2((m-l+1) + (m-l+2) + \cdots + m) \\ &= (2l+1)m - l^2.\end{aligned}$$

Thus, $M = \max\limits_{1 \leqslant i \leqslant m} \sum_{j=1}^{n} a_{i,j} \leqslant (2l+1)m - l^2$.

The minimal possible value of M is equal to

$$(2l+1)m-l^2=mn-\left(\frac{n-1}{2}\right)^2.$$

6 Prove that in an arithmetic progression consisting of 40 distinct positive integers, at least one of the numbers cannot be written as 2^k+3^l, where k, l are nonnegative integers. (Posed by Chen Yonggao)

Proof Suppose on the contrary that there exist 40 distinct positive integers in arithmetic progression such that each term can be written as 2^k+3^l, and denote this sequence by a, $a+d$, $a+2d$, ..., $a+39d$, where a, d are positive integers. Let

$$m=[\log_2(a+39d)],\ n=[\log_3(a+39d)].$$

In what follows, we first show that at most one of $a+26d$, $a+27d$, ..., $a+39d$ cannot be written as 2^m+3^l or 2^k+3^n (k, l are nonnegative integers).

Suppose that $a+hd$ cannot be written as 2^m+3^l or 2^k+3^n, for some $26\leqslant h\leqslant 39$. Then, by assumption, $a+hd=2^b+3^c$ for some nonnegative integers b, c. By the definition of m and n, it is clear that $b\leqslant m$, $c\leqslant n$. Since $a+hd$ cannot be written as 2^m+3^l or 2^k+3^n, we have $b\leqslant m-1$, $c\leqslant n-1$.

If $b\leqslant m-2$, then

$$\begin{aligned}a+hd&\leqslant 2^{m-2}+3^{n-1}=\frac{1}{4}\times 2^m+\frac{1}{3}\times 3^n\\&\leqslant\frac{7}{12}\times(a+39d)<a+26d,\end{aligned}$$

a contradiction.

If $c\leqslant n-2$, then

$$\begin{aligned}a+hd&\leqslant 2^{m-1}+3^{n-2}=\frac{1}{2}\times 2^m+\frac{1}{9}\times 3^n\\&\leqslant\frac{11}{18}\times(a+39d)<a+26d,\end{aligned}$$

also a contradiction.

It follows that $b = m - 1$, $c = n - 1$, which implies that at most one of $a + 26d$, $a + 27d$, ..., $a + 39d$ cannot be written as $2^m + 3^l$ or $2^k + 3^n$.

In these 14 numbers, at least 13 numbers can be written as $2^m + 3^l$ or $2^k + 3^n$. By the pigeonhole principle, at least 7 numbers can be written in the same form. We shall discuss two cases.

Case 1: There are 7 numbers in the form of $2^m + 3^l$, denoted by

$$2^m + 3^{l_1},\ 2^m + 3^{l_2},\ \ldots,\ 2^m + 3^{l_7},$$

where $l_1 < l_2 < \cdots < l_7$. Thus, 3^{l_1}, 3^{l_2}, ..., 3^{l_7} are the 7 terms of an arithmetic progression with 14 terms and the common difference d. However,

$$13d \geqslant 3^{l_7} - 3^{l_1} \geqslant \left(3^5 - \frac{1}{3}\right) \times 3^{l_2} > 13(3^{l_2} - 3^{l_1}) \geqslant 13d,$$

a contradiction.

Case 2: There are 7 numbers in the form of $2^k + 3^n$, denoted by

$$2^{k_1} + 3^n,\ 2^{k_2} + 3^n,\ \ldots,\ 2^{k_7} + 3^n,$$

where $k_1 < k_2 < \cdots < k_7$. Thus 2^{k_1}, 2^{k_2}, ..., 2^{k_7} are the 7 terms of an arithmetic progression with 14 terms and the common difference d. However,

$$13d \geqslant 2^{k_7} - 2^{k_1} \geqslant \left(2^5 - \frac{1}{2}\right) \times 2^{k_2} > 13(2^{k_2} - 2^{k_1}) \geqslant 13d,$$

a contradiction.

It follows from the above arguments that our assumption at the very beginning is false, which completes the proof.

2010 (Yingtan, Jiangxi)

First Day

(0800 - 1230; March 27, 2010)

1 For acute triangle ABC with $AB > AC$, let M be the midpoint of side BC and P a point inside $\triangle AMC$ such that $\angle MAB = \angle PAC$. Let O, O_1 and O_2 be the circumcenters of $\triangle ABC$, $\triangle ABP$ and $\triangle ACP$ respectively. Prove that line AO bisects segment O_1O_2. (Posed by Xiong Bin)

Proof I As shown in Fig. 1, draw the circumcircles of $\triangle ABC$, $\triangle ABP$ and $\triangle ACP$, respectively. Let the extension of AP meet $\odot O$ at D, join BD, and draw the line tangent to $\odot O$ at A, intersecting $\odot O_1$ and $\odot O_2$ at E and F respectively.

It is clear that $\triangle AMC \backsim \triangle ABD$, hence

$$\frac{AB}{BD} = \frac{AM}{MC}.$$

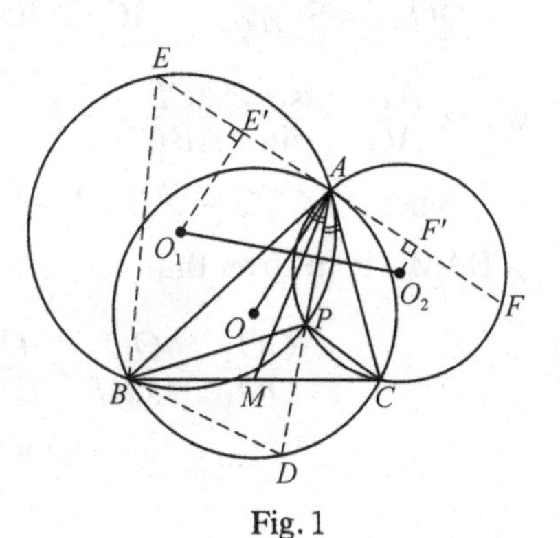

Fig. 1

Since $\triangle EAB \backsim \triangle PDB$, we have $\dfrac{AB}{BD} = \dfrac{AE}{PD}$.

Consequently, $\dfrac{AM}{MC} = \dfrac{AE}{PD}$, i.e.

$$AE = \frac{AM \times PD}{MC},$$

and, similarly,

$$AF = \frac{AM \times PD}{MB}.$$

It follows that

$$AE = AF. \quad ①$$

Draw the perpendicular lines $O_1E' \perp AE$ with foot E', and $O_2F' \perp AF$ with foot F'. Since E', F' are the midpoints of AE, AF respectively, it follows from ① that A is the midpoint of $E'F'$.

In the right-angled trapezoid $O_1E'F'O_2$, AO is the extension of the median, and hence it bisects the segment O_1O_2.

Proof II As shown in Fig. 2, draw segments AO_1, OO_1, AO_2, OO_2. Denote by Q the intersection of AO and O_1O_2. Then

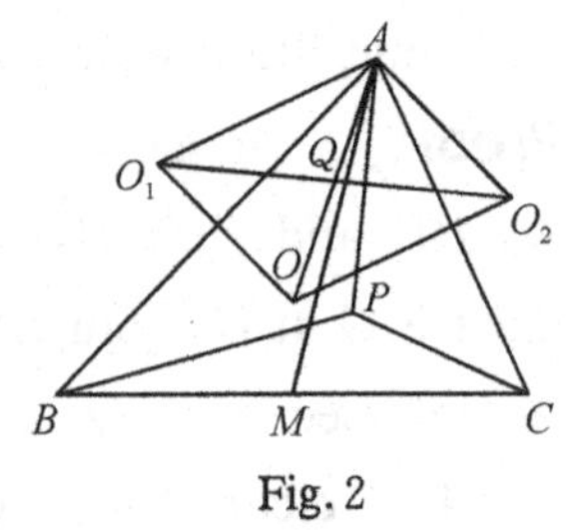

Fig. 2

$$\frac{O_1Q}{QO_2} = \frac{S_{\triangle AOO_1}}{S_{\triangle AOO_2}} = \frac{AB \times OO_1}{AC \times OO_2},$$

where $\frac{AB}{AC} = \frac{\sin\angle ACB}{\sin\angle ABC}$.

Since $\angle OO_1Q = \angle BAP = \angle CAM$, and $\angle OO_2Q = \angle CAP = \angle BAM$, it follows that

$$\begin{aligned}\frac{OO_1}{OO_2} &= \frac{OO_1}{OQ} \times \frac{OQ}{OO_2} \\ &= \frac{\sin\angle OQO_1}{\sin\angle OO_1Q} \times \frac{\sin\angle OO_2Q}{\sin\angle OQO_2} \\ &= \frac{\sin\angle OO_2Q}{\sin\angle OO_1Q} = \frac{\sin\angle BAM}{\sin\angle CAM},\end{aligned}$$

and thus

$$\frac{O_1Q}{QO_2} = \frac{\sin\angle ACM}{\sin\angle CAM} \times \frac{\sin\angle BAM}{\sin\angle ABM} = \frac{AM}{CM} \times \frac{BM}{AM} = \frac{BM}{CM}.$$

Note that M is the midpoint of BC, and therefore $O_1Q =$

QO_2, i.e. line AO bisects segment O_1O_2.

2 Let $A = \{a_1, a_2, \ldots, a_{2010}\}$ and $B = \{b_1, b_2, \ldots, b_{2010}\}$ be two sets of complex numbers, such that the equality

$$\sum_{1\leqslant i<j\leqslant 2010} (a_i + a_j)^n = \sum_{1\leqslant i<j\leqslant 2010} (b_i + b_j)^n$$

holds for every $n = 1, 2, \ldots, 2010$. Prove that $A = B$.
(Posed by Leng Gangsong)

Proof Let $S_k = \sum_{i=1}^{2010} a_i^k$ and $\widetilde{S}_k = \sum_{i=1}^{2010} b_i^k$. We first show by induction that $S_k = \widetilde{S}_k$ for $k = 1, 2, \ldots, 2010$.

Setting $n = 1$ in the given equality, we have $2009S_1 = 2009\widetilde{S}_1$, and hence $S_1 = \widetilde{S}_1$. Assume that $S_j = \widetilde{S}_j$ for $j = 1, 2, \ldots, k-1$, where $2 \leqslant k \leqslant 2010$; we are going to show that $S_k = \widetilde{S}_k$.

By the binomial expansion theorem, we have

$$\begin{aligned}
\sum_{1\leqslant i<j\leqslant 2010} (a_i + a_j)^k &= \sum_{1\leqslant i<j\leqslant 2010} \sum_{l=0}^{k} \binom{k}{l} a_i^l a_j^{k-l} \\
&= 2009S_k + \sum_{1\leqslant i<j\leqslant 2010} \sum_{l=0}^{k-1} \binom{k}{l} a_i^l a_j^{k-l} \\
&= 2009S_k + \frac{1}{2} \sum_{1\leqslant i\neq j\leqslant 2010} \sum_{l=1}^{k-1} \binom{k}{l} a_i^l a_j^{k-l} \\
&= 2009S_k + \frac{1}{2} \sum_{l=1}^{k-1} \sum_{i=1}^{2010} \binom{k}{l} a_i^l (S_{k-l} - a_i^{k-l}) \\
&= 2009S_k + \frac{1}{2} \sum_{l=1}^{k-1} \left(\binom{k}{l} S_{k-l} \sum_{i=1}^{2010} a_i^l - \sum_{i=1}^{2010} \binom{k}{l} a_i^k \right) \\
&= 2009S_k + \frac{1}{2} \sum_{l=1}^{k-1} \left(\binom{k}{l} S_{k-l} S_l - \binom{k}{l} S_k \right) \\
&= \frac{1}{2} \sum_{l=1}^{k-1} \binom{k}{l} S_{k-l} S_l + (2010 - 2^{k-1}) S_k. \qquad ①
\end{aligned}$$

Similarly, we have

$$\sum_{1\leqslant i<j\leqslant 2010}(b_i+b_j)^k=\frac{1}{2}\sum_{l=1}^{k-1}\binom{k}{l}\widetilde{S}_{k-l}\widetilde{S}_l+(2010-2^{k-1})\widetilde{S}_k. \quad ②$$

Since

$$\sum_{1\leqslant i<j\leqslant 2010}(a_i+a_j)^k=\sum_{1\leqslant i<j\leqslant 2010}(b_i+b_j)^k,$$

by ①, ② and inductive hypothesis $S_i=\widetilde{S}_i$ for $i=1, 2, \ldots, k-1$, we have $S_k=\widetilde{S}_k$ (it is worth noting that 2010 is not a power of 2, i.e. $2010-2^{k-1}\neq 0$). This completes the inductive proof that $S_k=\widetilde{S}_k$ for all $k=1, 2, \ldots, 2010$.

Set

$$(x-a_1)\cdots(x-a_{2010})=x^{2010}+A_1x^{2009}+\cdots+A_{2010}, \quad ③$$

$$(x-b_1)\cdots(x-b_{2010})=x^{2010}+B_1x^{2009}+\cdots+B_{2010}. \quad ④$$

By Newton's formula, we have

$$S_k+A_1S_{k-1}+\cdots+A_{k-1}S_1+kA_k=0, \quad ⑤$$

$$\widetilde{S}_k+B_1\widetilde{S}_{k-1}+\cdots+B_{k-1}\widetilde{S}_1+kB_k=0, \quad ⑥$$

for $k=1, 2, \ldots, 2010$.

It follows from ⑤, ⑥ and $S_k=\widetilde{S}_k$, $k=1, 2, \ldots, 2010$, by the easy inductive argument, we have

$$A_k=B_k,\ k=1, 2, \ldots, 2010.$$

The right hand sides of equations ③ and ④ are equal, and so are their left hand sides, i.e. $A=B$.

3 Let $n_1, n_2, \ldots, n_{26}$ be pairwise distinct positive integers, satisfying:

(1) In the decimal representation of each n_i, each digit

belongs to the set $\{1, 2\}$;

(2) For any i, j, n_j cannot be obtained from n_i by adding some digits on the right.

Find the least possible value of $\sum_{i=1}^{26} S(n_i)$, where $S(m)$ denotes the sum of all digits of m in decimal representation. (Posed by Fu Yunhao)

Solution Given two positive integers a, b in decimal representation, we say a *contains* b if a can be obtained from b by adding some digits on the right. We first prove a lemma.

Lemma Let $n_1, n_2, \ldots, n_r$ be pairwise distinct positive integers with digit 1 or 2. If none contains another, then the number of n_i's with $S(n_i) \leqslant t$ is at most F_t, where t is an arbitrary positive integer and F_t is a Fibonacci number satisfying $F_1 = 1$, $F_2 = 2$, $F_{n+2} = F_{n+1} + F_n (n \geqslant 1)$.

Proof of lemma. We induct on t. It is clear for $t = 1, 2$ (when $t = 2$, 1 and 11 cannot both appear). Suppose that the lemma is true for all positive integers less than t $(t \geqslant 3)$; we shall prove that it is also true for t. Suppose that without loss of generality $S(n_1), S(n_2), \ldots, S(n_l)$ are all the numbers with the sum of digits $\leqslant t$, where $n_1, n_2, \ldots, n_j$ start with 1 and $n_{j+1}, n_{j+2}, \ldots, n_l$ start with 2. If one of $n_1, n_2, \ldots, n_j$ is 1, then $j = 1 \leqslant F_{t-1}$, otherwise by deleting the first digit of n_1, $n_2, \ldots, n_j$ we obtain j positive integers with none containing another and the sum of digits $\leqslant t - 1$, and thus we again have $j \leqslant F_{t-1}$ by the inductive hypothesis. Analogously, we have $l - j \leqslant F_{t-2}$, and therefore $l \leqslant F_{t-1} + F_{t-2} = F_t$, i.e. the lemma is also true for t. The proof of the lemma is completed.

Going back to the original problem, we consider a more general question. Replacing 26 by m, denote the least possible value of $\sum_{i=1}^{m} S(n_i)$ by $f(m)$. Fix $m \geqslant 3$, and let $n_1, n_2, \ldots, n_m$ be a set of numbers satisfying the conditions in the problem which attains the minimum $f(m)$. Without loss of generality, assume that $\max_{1 \leqslant i \leqslant m} S(n_i) = S(n_1)$ and n_1 is maximum among all these numbers attaining the maximum digital sum. Since $m \geqslant 3$, n_1 contains at least two digits.

If the last digit of n_1 is 1, replace n_1 by $\frac{n_1 - 1}{10}$, which is not one of $n_2, n_3, \ldots, n_m$, otherwise n_1 would contain some n_i. Notice that $\frac{n_1 - 1}{10}, n_2, \ldots, n_m$ again satisfy the conditions in the problem, for if n_i contains $\frac{n_1 - 1}{10}$ for some $i \geqslant 2$, then $S(n_i) > S(n_1)$ and $n_i > n_1$, a contradiction. Now

$$S\left(\frac{n_1 - 1}{10}\right) + S(n_2) + \cdots + S(n_m) = f(m) - 1,$$

a contradiction to the definition of $f(m)$. Thus, the last digit of n_1 is 2.

If $n_1 - 1$ is not one of $n_2, n_3, \ldots, n_m$, replace n_1 by $n_1 - 1$, and these m numbers again satisfy the conditions in the problem, for if n_i contains $n_1 - 1$ for some $i \geqslant 2$, then n_i must be $10(n_1 - 1) + 1$, $S(n_i) = S(n_1)$; however, $n_i > n_1$ is a contradiction to the choice of n_1. Now

$$S\left(\frac{n_1 - 1}{10}\right) + S(n_2) + \cdots + S(n_m) = f(m) - 1,$$

a contradiction to the definition of $f(m)$. Thus, $n_1 - 1$ appears in $n_2, n_3, \ldots, n_m$.

Without loss of generality, assume that $n_2 = n_1 - 1$. Consider $\frac{n_1 - 2}{10}, n_3, \ldots, n_m$; since $\frac{n_1 - 2}{10} \neq n_i$ for $i \geqslant 3$, these are $m-1$ pairwise distinct numbers. There is no containment among n_2, ..., n_m, and $\frac{n_1-2}{10}$ does not contain any of n_3, ..., n_m, otherwise n_1 would contain that number. If one of n_3, ..., n_m contains $\frac{n_1-2}{10}$, say n_3, since $S\left(\frac{n_1-2}{10}\right)=S(n_1)-2$, n_3 is obtained by adding 1, 2 or 11 after $\frac{n_1-2}{10}$. Adding 1 or 2 yields n_2, n_1, and we must have $n_3=100\cdot\frac{n_1-2}{10}+11=10n_1-9$. Now $S(n_3)=S(n_1)$ and $n_3>n_1$, a contradiction to the choice of n_1. So $\frac{n_1-2}{10}$, n_3, ..., n_m satisfy the conditions in the problem, and therefore the sum of their digits is at least $f(m-1)$. Thus,

$$f(m)-S(n_1)-(S(n_1)-1)+(S(n_1)-2)\geqslant f(m-1),$$

i.e.

$$f(m)\geqslant f(m-1)+S(n_1)+1.$$

Let u be such that $F_{u-1}<m\leqslant F_u$. By the lemma, there are at most F_{u-1} of $S(n_1)$, $S(n_2)$, ..., $S(n_m)$ less than or equal to $u-1$, so $S(n_1)\geqslant u$, and

$$f(m)\geqslant f(m-1)+u+1. \qquad ①$$

It is easy to see that $f(1)=1$, $f(2)=3$, and hence

$$\begin{aligned}f(26)&=f(2)+\sum_{i=3}^{26}(f(i)-f(i-1))\\&=f(2)+(f(3)-f(2))+(f(5)-f(3))+(f(8)-f(5))\\&\quad+(f(13)-f(8))+(f(21)-f(13))+(f(26)-f(21))\\&\geqslant 3+4\times1+5\times2+6\times3+7\times5+8\times8+9\\&\quad\times5 \text{ (by equation ①)}\\&=179,\end{aligned}$$

i.e. $\sum_{i=1}^{25} S(n_i) \geqslant 179$.

On the other hand, by the property of Fibonacci numbers, there are exactly 8 numbers consisting of digits 1 and 2, with digital sum 5, denoted by $a_1, a_2, \ldots, a_8$, and there are exactly 13 such numbers with digital sum 6, denoted by $b_1, b_2, \ldots, b_{13}$. Add a digit 2 after each of $a_1, a_2, \ldots, a_8$, denoting these new numbers by $c_1, c_2, \ldots, c_8$. Add a digit 1 (resp. digit 2) to each of b_1, b_2, b_3, b_4, b_5, denoting these new numbers by d_1, d_2, d_3, d_4, d_5 (resp. e_1, e_2, e_3, e_4, e_5). Now consider

$$c_1, c_2, \ldots, c_8, d_1, d_2, \ldots, d_5, e_1, e_2, \ldots, e_5, b_6, b_7, \ldots, b_{13}.$$

These are pairwise distinct numbers consisting of digits 1 and 2, with total digital sum $7 \times 8 + 7 \times 5 + 8 \times 5 + 6 \times 8 = 179$. And there is none containing another. In fact, if x contains y, since their digital sum is either 6, 7 or 8, and a number with digital sum 8 ends up with 2, it follows that x has exactly one more digit than y. However, deleting the last digit of $d_1, d_2, \ldots, d_5$ and $e_1, e_2, \ldots, e_5$ yields $b_1, b_2, \ldots, b_5$, and deleting the last digit of $c_1, c_2, \ldots, c_8$ yields $a_1, a_2, \ldots, a_8$, none of which is in this set of 26 numbers.

We conclude that the least possible value of $\sum_{i=1}^{26} S(n_i)$ is 179.

Second Day

(0800 - 1230; March 28, 2010)

4 Let $G = G(V; E)$ be a simple graph with vertex set V and edge set E, and assume that $|V| = n$. A map $f: V \to \mathbf{Z}$ is said to be *good* if f satisfies:

(1) $\sum_{v \in V} f(v) = |E|$;

(2) If one colors arbitrarily some vertices into red, there exists a red vertex v, such that $f(v)$ is not greater than the number of vertices adjacent to v that are not colored into red.

Let $m(G)$ be the number of good maps f. Show that if each vertex of V is adjacent to a least one other vertex, then $n \leqslant m(G) \leqslant n!$. (Posed by Qu Zhenhua)

Proof Given an ordering $\tau = (v_1, v_2, \ldots, v_n)$ on the vertices in V, we associate a map $f_\tau : V \to \mathbf{Z}$ as follows: $f_\tau(v)$ is equal to the number of vertices in V that are ordered preceding v. We claim that f_τ is good.

Each edge is counted exactly once in $\sum_{v \in V} f_\tau(v)$, for an edge $e \in E$ with vertices $u, v \in V$ such that u is ordered before v in τ; then e is counted once in $f_\tau(v)$. Thus,

$$\sum_{v \in V} f_\tau(v) = |E|.$$

For any nonempty subset $A \subseteq V$ of all red vertices, choose $v \in A$ with the most preceding ordenings in τ. Then by definition, $f_\tau(v)$ is not greater than the number of vertices adjacent to v that are not colored into red. We have verified that f_τ is good.

Conversely, given any good map $f : V \to \mathbf{Z}$, we claim that $f = f_\tau$ for some ordering τ of V.

First, let the red vertice set $A = V$. By the condition (2) in the problem, there exists $v \in A$ such that $f(v) \leqslant 0$, and denote one of such vertices by v_1. Assuming that we have already chosen $v_1, \ldots, v_k$ from V, if $k < n$, set the red vertice set

$A = V - \{v_1, \ldots, v_k\}$. By the condition (2) in the problem, there exists $v \in A$ such that $f(v)$ is less than or equal to the number of vertices in $v_1, \ldots, v_k$ that are adjacent to v. Denote one of such vertices by v_{k+1}. Continuing in this way, we order the vertices by $\tau = (v_1, v_2, \ldots, v_n)$. By the construction we have $f(v) \leqslant f_\tau(v)$ for any $v \in V$. By the condition (1) in the problem, we have

$$|E| = \sum_{v \in V} f(v) \leqslant \sum_{v \in V} f_\tau(v) = |E|,$$

and therefore $f(v) = f_\tau(v)$ for any $v \in V$.

We have shown that for any ordering τ, f_τ is good, and any good map f is f_τ for some τ. Since the number of orderings on V is $n!$, we see that $m(G) \leqslant n!$ (note that two distinct orderings may result in the same map).

Next, we prove that $n \leqslant m(G)$. Assume at the moment that G is connected. Pick arbitrarily $v_1 \in V$. By the connectivity, we may choose $v_2 \in V - \{v_1\}$ such that v_2 is adjacent to v_1, and again we may choose $v_3 \in V - \{v_1, v_2\}$ such that v_3 is adjacent to at least one of v_1, v_2. Continuing in this way, we get an ordering $\tau = (v_1, v_2, \ldots, v_n)$ such that v_k is adjacent to at least one of the vertices preceding it under ordering τ, for any $2 \leqslant k \leqslant n$. Thus, $f_\tau(v_1) = 0$ and $f_\tau(v_k) > 0$ for $2 \leqslant k \leqslant n$. Since v_1 may be arbitrary, we have at least n good maps.

In general, if G is a union of its connected components $G_1, \ldots, G_k$, since each vertex is adjacent to at least another vertex, each component has at least two vertices, denote by $n_1, \ldots, n_k \geqslant 2$ the number of vertices of these components. For each G_i, we have at least n_i good maps on its vertices, $i = 1, \ldots, k$. It is easy to see that patching good maps on G_i's together results in a good map on G, and thus

$$m(G) \geqslant n_1 n_2 \cdots n_k \geqslant n_1 + n_2 + \cdots + n_k = n.$$

We conclude that $n \leqslant m(G) \leqslant n!$.

5 Given integer $a_1 \geqslant 2$, for $n \geqslant 2$, define a_n to be the least positive integer not coprime to a_{n-1} and not equal to a_1, a_2, ..., a_{n-1}. Prove that every integer except 1 appears in the sequence $\{a_n\}$. (Posed by Yu Hongbing)

Proof We proceed in three steps:

Step 1: We prove that there are infinitely many even numbers in this sequence.

Suppose on the contrary that there are only finitely many even numbers, and there is an integer E, such that all even numbers greater than E do not appear in the sequence. It follows that there exists a positive integer K such that a_n is an odd number greater than E for any $n \geqslant K$. Then there is some $n_1 > K$ such that $a_{n_1+1} > a_{n_1}$ (otherwise the sequence is strictly decreasing after a_{n_1}, a contradiction).

Let p be the smallest prime divisor of a_{n_1}, $p \geqslant 3$. Since

$$(a_{n_1+1} - a_{n_1}, a_{n_1}) = (a_{n_1+1}, a_{n_1}) > 1,$$

we have $a_{n_1+1} - a_{n_1} \geqslant p$, i.e. $a_{n_1+1} \geqslant a_{n_1} + p$.

On the other hand, $a_{n_1} + p$ is even and greater than E, so it does not appear before a_{n_1}, and therefore $a_{n_1+1} = a_{n_1} + p$, which is even — a contradiction.

Step 2: We prove that all even numbers are in this sequence.

Suppose on the contrary that $2k$ is not in this sequence and is the smallest such even number. Let $\{a_{n_i}\}$ be the subsequence of $\{a_n\}$ consisting of all even numbers. By step 1, it is an infinite sequence. Since $(a_{n_i}, 2k) > 1$ and $2k \notin \{a_n\}$, we have $a_{n_i+1} \leqslant 2k$ by definition.

However, $\{a_{n_1+1}\}$ is infinite — a contradiction. Thus, $\{a_n\}$ contains all even numbers.

Step 3: We prove that $\{a_n\}$ contains all odd numbers greater than 1.

Suppose on the contrary that $2k+1$ is an odd integer greater than 1 which is not in $\{a_n\}$, and is the smallest such number. By step 2, there is an infinite subsequence $\{a_{m_i}\}$ of $\{a_n\}$ consisting of even numbers that are multiples of $2k+1$. Arguing analogously as in step 2, we have $a_{m_i+1} \leqslant 2k+1$, $i = 1, 2, \ldots$, a contradiction.

We have shown that $\{a_n\}$ contains all positive integers except 1.

6 Given integer $n \geqslant 2$ and real numbers $x_1, x_2, \ldots, x_n$ in the interval $[0, 1]$, prove that there exist real numbers a_0, $a_1, \ldots, a_n$ satisfying simultaneously the following conditions:

(1) $a_0 + a_n = 0$;

(2) $|a_i| \leqslant 1$, for every $i = 0, 1, \ldots, n$;

(3) $|a_i - a_{i-1}| = x_i$, for every $i = 1, 2, \ldots, n$.

(Posed by Zhu Huawei)

Proof For any $a \in [0, 1)$, define a sequence $\{a_i\}_{i=0}^{n}$ generated by a as follows: $a_0 = a$ for $1 \leqslant i \leqslant n$, $a_i = a_{i-1} - x_i$ if $a_{i-1} \geqslant 0$, and $a_i = a_{i-1} + x_i$ if $a_{i-1} < 0$.

Set $f(a) = a_n$. It is easy to show by induction that $|a_i| \leqslant 1$ for every $0 \leqslant i \leqslant n$.

If there exists $a \in [0, 1)$ such that $f(a) = -a$, consider the sequence generated by a, $a_0 = a, a_1, \ldots, a_n = f(a) = -a$. Clearly, this sequence satisfies the conditions (1) and (3). By the recursive relation, it also satisfies the condition (2). Thus, it suffices to show that there exists $a \in [0, 1)$ such that $f(a) = -a$.

For any $a \in [0, 1)$, we say that a is a breaking point if at least one term in its generating sequence is 0. Since every breaking point is of the form $\sum_{i=1}^{n} t_i x_i$, where $t_i = -1, 0, 1$, there are finitely many breaking points.

Clearly, 0 is a breaking point; label all breaking points in increasing order by $0 = b_1 < b_2 < \cdots < b_m < 1$.

We first prove that for $1 \leqslant k \leqslant m-1$, $f(a) = f(b_k) + (a - b_k)$ for every $a \in [b_k, b_{k+1})$.

Consider b_k, b_{k+1} and their generating sequences. Assume that $q_0 = b_k, q_1, q_2, \ldots, q_n$ is the generating sequence of b_k, and $r_0 = b_{k+1}, r_1, r_2, \ldots, r_n$ is the generating sequence of b_{k+1}. Suppose that r_l is the first term of $\{r_i\}_{i=0}^{n}$ equal to 0. Construct a sequence $\{s_i\}_{i=0}^{n}$ as follows:

$$s_0 = r_0, s_1 = r_1, \ldots, s_l = r_l = 0,$$

$$s_{l+1} = -r_{l+1}, \ldots, s_n = -r_n.$$

It is clear that the sequence $\{s_i\}_{i=0}^{n}$ satisfies $s_0 = b_{k+1}$; and for $1 \leqslant i \leqslant n$, $s_i = s_{i-1} - x_i$ if $s_{i-1} > 0$, and $s_i = s_{i-1} + x_i$ if $s_{i-1} \leqslant 0$.

We prove by induction that $q_i s_i \geqslant 0$ and $s_i - q_i = b_{k+1} - b_k$. The conclusion is obviously true for $i = 0$. Assume that it holds for $i - 1$. Then $q_{i-1} s_{i-1} \geqslant 0$ and $s_{i-1} - q_{i-1} = b_{k+1} - b_k > 0$, which implies that $q_{i-1} \geqslant 0$, $s_{i-1} > 0$, or $q_{i-1} < 0$, $s_{i-1} \leqslant 0$. In the former case, $q_i = q_{i-1} - x_i$, $s_i = s_{i-1} - x_i$, and thus $s_i - q_i =$

$s_{i-1} - q_{i-1} = b_{k+1} - b_k$. Similarly, in the latter case, we have $s_i - q_i = b_{k+1} - b_k$. If $q_i s_i < 0$, then $q_i < 0 < s_i$. Set $b' = b_{k+1} - s_i = b_k + (-q_i) \in (b_k, b_{k+1})$, and consider the generating sequence of b', $u_0 = b'$, u_1, ..., u_n. It is easy to show by induction that $s_j - u_j = s_0 - u_0 = s_i$ for any $0 \leqslant j \leqslant i$ (q_{j-1} and s_{j-1} both add or subtract x_j for $j < i$; u_j lies between q_{j-1} and s_{j-1}, so the recursive relation is the same). Then $u_i = s_i - s_i = 0$, i.e. b' is a breaking point — a contradiction to b_k, b_{k+1} being two consecutive breaking points. Therefore, $q_i s_i \geqslant 0$. By induction we have verified that $q_i s_i \geqslant 0$ and $s_i - q_i = b_{k+1} - b_k$ for all $0 \leqslant i \leqslant n$.

Since $f(b_k) = q_n$, $f(b_{k+1}) = r_n = -s_n$, we have $f(b_k) + f(b_{k+1}) = b_k - b_{k+1}$. It follows from the above argument that, for any $b_k < b' < b_{k+1}$, the generating sequence of b' has the same recursive relation as $\{q_i\}_{i=0}^n$ and $\{s_i\}_{i=0}^n$, and hence

$$f(b') = f(b_k) + (b' - b_k).$$

Let us go back to the original problem.

If $f(b_k) = -b_k$ for some k, we are done. If $f(b_k) = b_k$ for some k, then consider the generating sequence of b_k, reversing every term after the first 0, and we obtain a new sequence $z_0 = b_k$, z_1, ..., $z_n = -b_k$, which satisfies the required conditions. Now assume that $|f(b_k)| \neq b_k$ for every k, and we shall consider two cases to show that $f(a) = -a$ for some $a \in [0, 1)$:

Case 1: $|f(b_m)| < b_m$. Since $|f(b_1)| > b_1 = 0$, there exists k such that $|f(b_k)| > b_k$, $|f(b_{k+1})| < b_{k+1}$.

Since $f(b_k) + f(b_{k+1}) = b_k - b_{k+1}$, we have $f(b_k) - b_k = -(f(b_{k+1}) + b_{k+1}) < 0$, and hence $f(b_k) < -b_k$. Again by $f(b_k) + f(b_{k+1}) = b_k - b_{k+1}$, we have

$$f(b_k) = b_k - b_{k+1} - f(b_{k+1}) > b_k - 2b_{k+1},$$

i.e.

$$b_k - 2b_{k+1} < f(b_k) < -b_k.$$

Let $b' = \dfrac{b_k - f(b_k)}{2}$. Then $b_k < b' < b_{k+1}$, and

$$f(b') = f(b_k) + (b' - b_k) = (b_k - 2b') + (b' - b_k) = -b',$$

and the result follows.

Case 2: $|f(b_m)| > b_m$. Since there is no breaking number in $(b_m, 1)$, we see from the previous argument that for any $b' \in (b_m, 1)$, the generating sequence of b' and the generating sequence of b_m have the same recursive relation, and hence

$$f(b') = f(b_m) + (b' - b_m).$$

Since $|f| \leqslant 1$, we have $f(b_m) \leqslant b_m$, and hence

$$-1 \leqslant f(b_m) < -b_m.$$

Let $b' = \dfrac{b_m - f(b_m)}{2}$. Then

$$b_m = \frac{b_m - (-b_m)}{2} < b' < \frac{b_m - (-1)}{2} = \frac{b_m + 1}{2} < 1,$$

$$\begin{aligned} f(b') &= f(b_m) + (b' - b_m) \\ &= (b_m - 2b') + (b' - b_m) = -b'. \end{aligned}$$

The result again follows.

China Girls' Mathematical Olympiad

2008 (Zhongshan, Guangdong)

1 (1) Can one divide the set $\{1, 2, \ldots, 96\}$ into 32 subsets, each containing three elements, and the sums of the three elements in each subset are all equal?

(2) Can one divide the set $\{1, 2, \ldots, 99\}$ into 33 subsets, each containing three elements, and the sums of the three elements in each subset are all equal? (Posed by Liu Shixiong)

Solution (1) No. As

$$1+2+\cdots+96=\frac{96\times(96+1)}{2}=48\times 97,$$

and $32 \nmid 48 \times 97$.

(2) Yes. The sum of the three elements in each set is

$$\frac{1+2+\cdots+99}{33} = \frac{99 \times (99+1)}{33 \times 2} = 150.$$

We can divide 1, 2, 3, ..., 66 into 33 pairs, such that the sums of these pairs form an arithmetic sequence:

$$1+50,\ 3+49,\ \ldots,\ 33+34,\ 2+66,\ 4+65,\ \ldots,\ 32+51.$$

Hence, the following decomposition satisfies the requirement:

$\{1, 50, 99\}$, $\{3, 49, 98\}$, ..., $\{33, 34, 83\}$, $\{2, 66, 82\}$, $\{4, 65, 81\}$, ..., $\{32, 51, 67\}$.

Remark The general case is as follows:

Let the family of subsets $A_i = \{x_i, y_i, z_i\}$ of the set $M = \{1, 2, 3, \ldots, 3n\}$, $i = 1, 2, \ldots, n$, satisfy $A_1 \cup A_2 \cup \cdots \cup A_n = M$. Write $s_i = x_i + y_i + z_i$, and find all possible values of n, such that the s_i' s are all equal.

This can be answered as follows: First, $n \mid 1+2+3+\cdots+3n$, i.e.

$$n \left| \frac{3n(3n+1)}{2} \right. \Rightarrow 2 \mid 3n+1.$$

Hence, n must be odd.

For n odd, 1, 2, 3, ..., $2n$ can form n pairs such that the sums of each pair become an arithmetic sequence of common difference 1:

$$1+\left(n+\frac{n+1}{2}\right),\ 3+\left(n+\frac{n-1}{2}\right),\ \ldots,\ n+(n+1);$$

$$2+2n,\ 4+(2n-1),\ \ldots,\ (n-1)+\left(n+\frac{n+3}{2}\right).$$

Its general term is

$$a_k = \begin{cases} 2k-1+\left(n+\dfrac{n+1}{2}+1-k\right), & 1 \leqslant k \leqslant \dfrac{n+1}{2}, \\ [1-n+2(k-1)]+\left[2n+\dfrac{n+1}{2}-(k-1)\right], & \dfrac{n+3}{2} \leqslant k \leqslant n. \end{cases}$$

It is easy to see that $a_k + 3n + 1 - k = \dfrac{9n+3}{2}$ is a constant, so all the following n triples have the same sum:

$$\left\{1, n+\frac{n+1}{2}, 3n\right\}, \left\{3, n+\frac{n-1}{2}, 3n-1\right\}, \ldots,$$
$$\left\{n, n+1, 3n+1-\frac{n+1}{2}\right\};$$
$$\left\{2, 2n, 3n+1-\frac{n+3}{2}\right\}, \ldots, \left\{n-1, n+\frac{n+3}{2}, 2n+1\right\}.$$

For n odd, take these triples as $A_1, A_2, \ldots, A_n$; then they satisfy the required condition. So n can be any odd number.

2 The real polynomial $\varphi(x) = ax^3 + bx^2 + cx + d$ has three positive roots, and $\varphi(0) < 0$. Prove that

$$2b^3 + 9a^2d - 7abc \leqslant 0. \quad ①$$

(Posed by Zhu Huawei)

Proof We denote by x_1, x_2, x_3 the three positive roots of the polynomial $\varphi(x) = ax^3 + bx^2 + cx + d$. By Vieta's theorem, we have

$$x_1 + x_2 + x_3 = -\frac{b}{a},$$

$$x_1x_2 + x_2x_3 + x_3x_1 = \frac{c}{a},$$

$$x_1x_2x_3 = -\frac{d}{a}.$$

As $\varphi(0) < 0$, we get $d < 0$, and hence $a > 0$.

Dividing both sides of ① by a^3, we get an equivalent form of ①:

$$7\left(-\frac{b}{a}\right)\frac{c}{a} \leqslant 2\left(-\frac{b}{a}\right)^3 + 9\left(-\frac{d}{a}\right)$$

$$\Leftrightarrow 7(x_1 + x_2 + x_3)(x_1x_2 + x_2x_3 + x_3x_1) \leqslant 2(x_1 + x_2 + x_3)^3 + 9x_1x_2x_3$$

$$\Leftrightarrow x_1^2x_2 + x_1^2x_3 + x_2^2x_1 + x_2^2x_3 + x_3^2x_1 + x_3^2x_2 \leqslant 2(x_1^3 + x_2^3 + x_3^3). \qquad ②$$

Because x_1, x_2, x_3 are all greater than 0, $(x_1 - x_2)(x_1^2 - x_2^2) \geqslant 0$. That is to say,

$$x_1^2x_2 + x_2^2x_1 \leqslant x_1^3 + x_2^3.$$

By the same argument, $x_2^2x_3 + x_3^2x_2 \leqslant x_2^3 + x_3^3$, $x_3^2x_1 + x_1^2x_3 \leqslant x_3^3 + x_1^3$.

Summing up these three inequalities we get ②, and the equality holds iff $x_1 = x_2 = x_3$.

3 Find the smallest constant $a > 1$, such that for any point P inside a square $ABCD$ there exist two triangles among $\triangle PAB$, $\triangle PBC$, $\triangle PCD$, $\triangle PDA$, with the ratio between their areas belonging to the interval $[a^{-1}, a]$. (Posed by Li Weigu)

Solution $a_{\min} = \frac{1+\sqrt{5}}{2}$. We first prove that $a_{\min} \leqslant \frac{1+\sqrt{5}}{2}$.

Write $\varphi = \frac{1+\sqrt{5}}{2}$. We may assume that each edge has length $\sqrt{2}$. For any point P inside the square $ABCD$, let S_1, S_2, S_3, S_4 denote the area of $\triangle PAB$, $\triangle PBC$, $\triangle PCD$, $\triangle PDA$ respectively; we can also assume that $S_1 \geqslant S_2 \geqslant S_4 \geqslant S_3$.

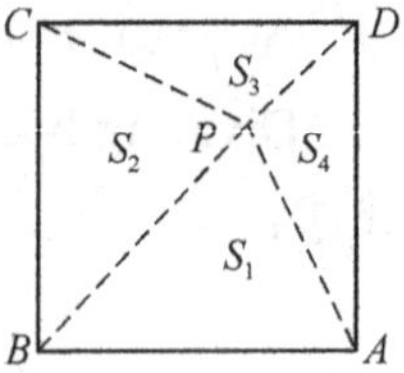

Let $\lambda = \frac{S_1}{S_2}$, $\mu = \frac{S_2}{S_4}$. If λ, $\mu > \varphi$, as

$$S_1 + S_3 = S_2 + S_4 = 1,$$

we have $\frac{S_2}{1 - S_2} = \mu$, $S_2 = \frac{\mu}{1+\mu}$. So

$$S_1 = \lambda S_2 = \frac{\lambda\mu}{1+\mu} = \frac{\lambda}{1+\frac{1}{\mu}} > \frac{\varphi}{1+\frac{1}{\varphi}} = \frac{\varphi^2}{1+\varphi} = 1,$$

and we reach a contradiction. Hence, $\min\{\lambda, \mu\} \leqslant \varphi$, which implies that $a_{\min} \leqslant \varphi$.

On the other hand, for any $a \in (1, \varphi)$, we take any $t \in \left(a, \frac{1+\sqrt{5}}{2}\right)$ such that $b = \frac{t^2}{1+t} > \frac{8}{9}$. Inside the square $ABCD$ we can choose a point P so that $S_1 = b$, $S_2 = \frac{b}{t}$, $S_3 = \frac{b}{t^2}$, $S_4 = 1 - b$. Then we have

$$\frac{S_1}{S_2} = \frac{S_2}{S_3} = t \in \left(a, \frac{1+\sqrt{5}}{2}\right),$$

$$\frac{S_3}{S_4} = \frac{b}{t^2(1-b)} > \frac{b}{4(1-b)} > 2 > a.$$

Thus, for any $i, j \in \{1, 2, 3, 4\}$, we have $\frac{S_i}{S_j} \notin [a^{-1}, a]$.

Hence $a_{\min} = \varphi$.

4 Outside a convex quadrilateral $ABCD$ we construct equilateral triangles ABQ, BCR, CDS and DAP. Denoting by x the sum of the diagonals of $ABCD$, and by y the sum of line segments joining the midpoints of opposite sides of $PQRS$, we find the maximum value of $\frac{y}{x}$. (Posed by Xiong Bin)

Solution If $ABCD$ is a square, then $\frac{y}{x}=\frac{1+\sqrt{3}}{2}$.

Now we prove that $\frac{y}{x} \leqslant \frac{1+\sqrt{3}}{2}$.

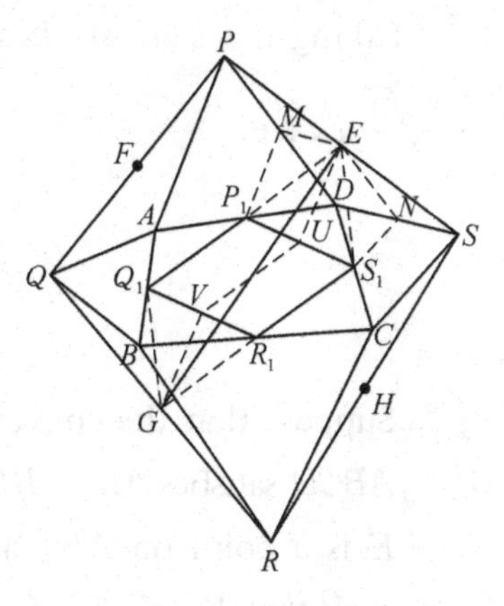

Denote by P_1, Q_1, R_1, S_1 the midpoints of DA, AB, BC, CD, and by E, F, G, H the midpoints of SP, PQ, QR, RS. Then $P_1Q_1R_1S_1$ is a parallelogram.

Now draw lines P_1E, S_1E, and denote by M, N the midpoints of DP, DS. Then

$$DS_1 = S_1N = DN = EM,$$
$$DP_1 = P_1M = MD = EN,$$

and

$$\begin{aligned}\angle P_1DS_1 &= 360^\circ - 60^\circ - 60^\circ - \angle PDS \\ &= 240^\circ - (180^\circ - \angle END) = 60^\circ + \angle END \\ &= \angle ENS_1 = \angle EMP_1.\end{aligned}$$

So we have $\triangle DP_1S_1 \cong \triangle MP_1E \cong \triangle NES_1$. Hence, $\triangle EP_1S_1$ is equilateral.

By the same argument, $\triangle GQ_1R_1$ is also equilateral. Now let U, V be the midpoints of P_1S_1, Q_1R_1, respectively. We then obtain

$$EG \leqslant EU + UV + VG = \frac{\sqrt{3}}{2}P_1S_1 + P_1Q_1 + \frac{\sqrt{3}}{2}Q_1R_1$$

$$= P_1Q_1 + \sqrt{3}P_1S_1 = \frac{1}{2}BD + \frac{\sqrt{3}}{2}AC,$$

and also
$$FH \leqslant \frac{1}{2}AC + \frac{\sqrt{3}}{2}BD.$$

Taking the sum of these two inequalities, we have $y \leqslant \frac{1+\sqrt{3}}{2}x$, i.e.

$$\frac{y}{x} \leqslant \frac{1+\sqrt{3}}{2}.$$

5 Suppose that the convex quadrilateral $ABCD$ satisfies $AB = BC$, $AD = DC$. E is a point on AB, and F on AD, such that B, E, F, D are concyclic. Draw $\triangle DPE$ directly similar to $\triangle ADC$, and $\triangle BQF$ directly similar to $\triangle ABC$. Prove that A, P, Q are collinear. (Posed by Ye Zhonghao)

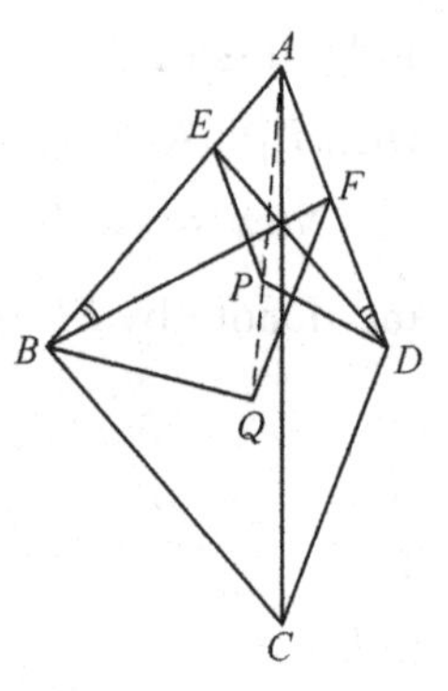

Fig. 1

Proof Denote by O the center of the circle that passes through B, E, F, D. Draw lines OB, OF, BD.

In $\triangle BDF$, O is the circumcenter, so $\angle BOF = 2\angle BDA$; And $\triangle ABD \backsim \triangle CBD$, so $\angle CDA = 2\angle BDA$. Hence, $\angle BOF = \angle CDA = \angle EPD$, which implies that the isosceles triangles

$$\triangle BOF \backsim \triangle EPD. \quad ①$$

On the other hand, the concyclicity of B, E, F, D implies that

$$\triangle ABF \backsim \triangle ADE. \quad ②$$

Combining ① and ②, we know that the quadrilateral $ABOF \backsim ADPE$, so

$$\angle BAO = \angle DAP. \qquad ③$$

The same argument gives

$$\angle BAO = \angle DAQ. \qquad ④$$

③ and ④ imply that A, P, Q are collinear.

Remark In fact, when $ABCD$ is not a rhombus, the collinearity of A, P, Q is equivalent to the concyclicity of B, E, F, D.

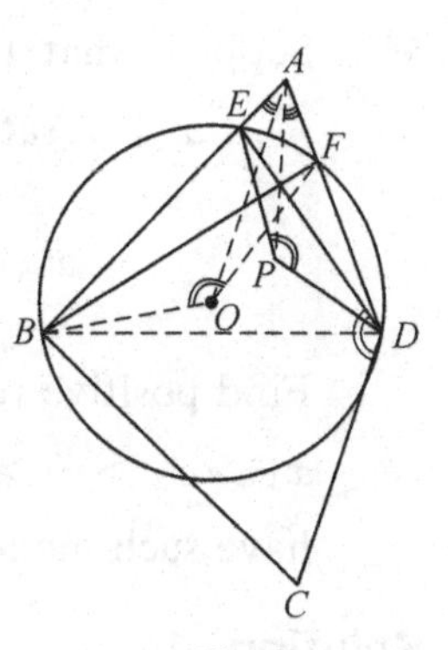

Fig. 2

This can be explained in the following way. Fix the point E, and let the point F move along the line AD. By similarity, the locus of Q is a line that passes P. Now it suffices to show that this locus does not coincide with the line AP, i.e. to show that A is not on the locus.

To this end, we draw $\triangle BAA' \backsim \triangle BQF \backsim \triangle ABC$ (see Fig. 3). Then $\angle BAA' = \angle ABC$, which implies that $A'A \parallel BC$. As $ABCD$ is not a rhombus, AD is not parallel to BC, which implies that A', A', D are not collinear, i.e. A is not on

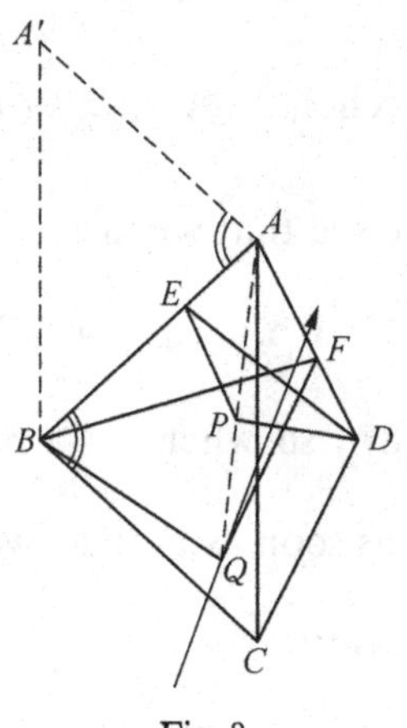

Fig. 3

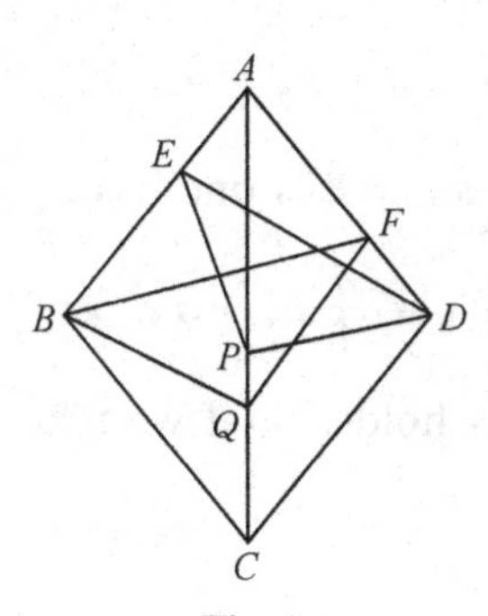

Fig. 4

the locus. Hence, Q is on the line AP only when B, E, F, D are concyclic. And when $ABCD$ is a rhombus (see Fig. 4), for any E and F the corresponding points P and Q are always on the diagonal AC.

6 Suppose that the sequence of positive numbers x_1, x_2, $\ldots$, x_n, $\ldots$ satisfies $(8x_2 - 7x_1)x_1^7 = 8$ and

$$x_{k+1}x_{k-1} - x_k^2 = \frac{x_{k-1}^8 - x_k^8}{(x_k x_{k-1})^7},\ k \geqslant 2.$$

Find positive real number a such that when $x_1 > a$ one has $x_1 > x_2 > \cdots > x_n > \cdots$, and when $0 < x_1 < a$ one does not have such monotonicity. (Posed by Li Shenghong)

Solution By $x_{k+1}x_{k-1} - x_k^2 = \frac{x_{k-1}^8 - x_k^8}{(x_k x_{k-1})^7}$, we have

$$\frac{x_{k+1}}{x_k} - \frac{x_k}{x_{k-1}} = \frac{1}{x_k^8} - \frac{1}{x_{k-1}^8},$$

i.e.

$$\frac{x_{k+1}}{x_k} - \frac{1}{x_k^8} = \frac{x_k}{x_{k-1}} - \frac{1}{x_{k-1}^8} = \cdots = \frac{x_2}{x_1} - \frac{1}{x_1^8} = \frac{7}{8}.$$

Hence, $x_{k+1} = \frac{7}{8}x_k + x_k^{-7}$, and when $x_1 > 0$, $x_k > 0$, $k \geqslant 2$.

By $x_{k+1} - x_k = x_k\left(x_k^{-8} - \frac{1}{8}\right)$, we see that when $x_k^{-8} - \frac{1}{8} < 0$, i.e. $x_k > 8^{\frac{1}{8}}$, one has $x_{k+1} - x_k < 0$, i.e. $x_{k+1} < x_k$, $k \geqslant 1$.

And $x_{k+1} = \frac{7}{8}x_k + x_k^{-7} \geqslant 8\sqrt[8]{\frac{1}{8^7}} = 8^{\frac{1}{8}}$, so when $x_k = 8^{\frac{1}{8}}$, the equality holds. So if we take $a = 8^{\frac{1}{8}}$, as soon as $x_k > 8^{\frac{1}{8}}$ we have

$$x_1 > x_2 > \cdots > x_n \cdots.$$

When $x_1 < 8^{\frac{1}{8}}$, we have $x_2 > x_1$ and $x_2 > x_3 > \cdots > x_n \cdots$.

So the constant that we are looking for is $a=8^{\frac{1}{8}}$.

7 For a chessboard of the size 2008×2008, in each case (they all have different colors) write one of the letters C, G, M, O. If every 2×2 square contains all these four letters, we call it a "harmonic chessboard." How many different harmonic chessboard are there? (Posed by Zuming Feng)

Solution There are $12\times 2^{2008}-24$ harmonic chessboards. We first prove the following claim:

In every harmonic chessboard, at least one of the following two situations occurs: (1) each line is composed of just two letters, in an alternative way; (2) each column is composed of just two letters, in an alternative way.

In fact, suppose that one line is not composed of two letters; then there must be three consecutive squares containing different letters. Without loss of generality, we may assume these three letters to be C, G, M, as shown in Fig. 1. We then get easily $X_2=X_5=O$, $X_1=X_4=M$ and $X_3=X_6=C$, as shown in Fig. 2.

X_1	X_2	X_3
C	G	M
X_4	X_5	X_6

Fig. 1

M	O	C
C	G	M
M	O	C

Fig. 2

The same argument shows that each of these three columns is composed of two letters in an alternative way, and *a fortiori* so is every column.

Now we calculate the total number of different harmonic chessboards. If the leftmost column is composed of two letters (eg. C and M), we see immediately that all the odd-numbered columns are composed of these two letters, while the even-numbered columns are composed of the other two letters. The letter in the top square of each column can be either of the two letters that compose this column; we check easily that it is a harmonic chessboard. Therefore, we have $\binom{4}{2} = 6$ different ways to choose the two letters of the first column, and 2^{2008} ways to determine the letter in the top square of each column. Hence, we get 6×2^{2008} configurations to make each column composed of two letters in an alternative way. We have also 6×2^{2008} configurations to make each line composed of two letters in an alternative way.

Then we need to subtract from the sum the configurations that are counted twice, i. e. the configurations that are alternative on each line and each column. Obviously, any such configuration is in one-to-one correspondence to the 2×2 square at the upper-left corner, which gives $4! = 24$ different ways. Hence, we get the above result.

8 For positive integer n, let $f_n = [2^n \sqrt{2008}] + [2^n \sqrt{2009}]$. Prove that there are infinitely many odd numbers and even numbers in the sequence $f_1, f_2, \ldots$. ($[x]$ represents the biggest integer that does not exceed x.) (Posed by Zuming Feng)

Proof We use the dyadic representations of $\sqrt{2008}$ and $\sqrt{2009}$:

$$\sqrt{2008} = \overline{101100.a_1a_2\cdots}_{(2)}, \quad \sqrt{2009} = \overline{101100.b_1b_2\cdots}_{(2)}.$$

First, we prove that there are infinitely many even numbers by contradiction. Suppose that there are only finitely many even numbers in the sequence. Then there exists a positive integer N, and for every positive integer $n > N$, f_n must be odd. We consider $n_1 = N + 1$, $n_2 = N + 2, \ldots$. We observe that in dyadic representation

$$f_{n_i} = \overline{101100b_1b_2\cdots b_{n_i}}_{(2)} + \overline{101100a_1a_2\cdots a_{n_i}}_{(2)}.$$

This number is equal to $b_{n_i} + a_{n_i}$ modulo 2. As f_{n_i} is odd, we have $\{b_{n_i}, a_{n_i}\} = \{0, 1\}$. Hence,

$$\sqrt{2008} + \sqrt{2009} = \overline{1011001.c_1c_2\cdots c_{N-1}111\cdots}_{(2)}.$$

Therefore, $\sqrt{2008} + \sqrt{2009}$ must be rational, which is impossible, as we know that it is irrational. Hence, our hypothesis must be wrong, which proves the existence of infinitely many even numbers in the sequence.

In a similar way we can prove the existence of infinitely many odd numbers in the sequence. Let

$$g_n = [2^n\sqrt{2009}] - [2^n\sqrt{2008}],$$

apparently g_n and f_n have the same parity. Hence, for $n > N$, g_n is even. We observe also that in dyadic representation

$$g_{n_i} = \overline{101100b_1b_2\cdots b_{n_i}}_{(2)} - \overline{101100a_1a_2\cdots a_{n_i}}_{(2)}.$$

This number is equal to $b_{n_i} - a_{n_i}$ modulo 2. As g_{n_i} is odd, we have $b_{n_i} = a_{n_i}$. Thus,

$$\sqrt{2009} - \sqrt{2008} = \overline{0.d_1d_2\cdots d_{N-1}000\cdots}_{(2)},$$

and this would imply the rationality of $\sqrt{2009} - \sqrt{2008}$, which is impossible. Hence, there are infinitely many odd numbers in the sequence.

2009 (Xiamen, Fujian)

First Day

(0800 - 1200; August 13, 2009)

1 Show that there are only finitely many triples (a, b, c) of positive integers satisfying the equation $abc = 2009(a+b+c)$. (Posed by Ieng Tak Leong)

Solution There are at most six permutations for any three numbers x, y, z. It suffices to show that there are only finitely many triples (a, b, c), with $a \geqslant b \geqslant c$, of positive integers satisfying the equation $abc = 2009(a+b+c)$. It follows that $abc \leqslant 2009 \times (3a)$ or $bc \leqslant 2009 \times 3 = 6027$. Clearly, there are finitely many pairs (b, c) of positive integers satisfying the equation $bc \leqslant 6027$ and for each fixed pair of integers (b, c) there is at most one positive integer a satisfying the equation $abc = 2009(a+b+c)$ (because it is a linear equation in a).

2 A right triangle ABC, with $\angle BAC = 90°$, is inscribed in the circle Γ. The point E lies in the interior of the arc $\overset{\frown}{BC}$ (not containing A), with $EA > EC$. The point F lies on the ray EC with $\angle EAC = \angle CAF$. The segment BF meets Γ again at D (other than B). Let O denote the circumcenter of the triangle DEF. Prove that the points A, C, O are collinear. (Posed by Bian Hongping)

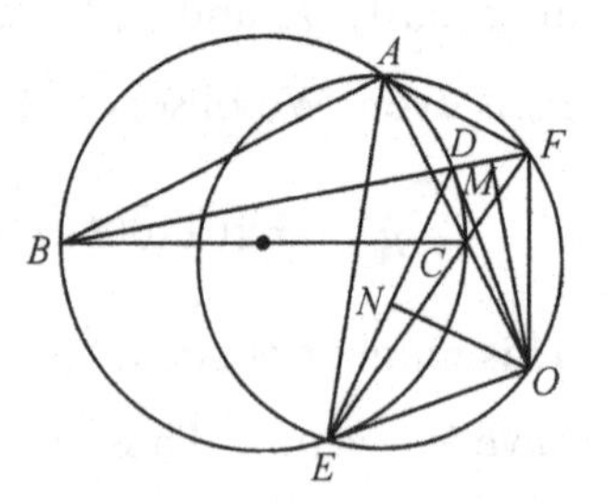

Solution Let M and N be the feet of the perpendiculars from

O to the lines DF and DE, respectively. Because O is the circumcenter of the triangle DEF, the triangles EOD and ODF are both isosceles with $EO = DO = FO$. It follows that

$$\angle EOF = \angle EOD + \angle DOF = 2\angle NOD + 2\angle DOM = 2\angle NOM.$$

Because $\angle OND = \angle OMD = 90°$, the quadrilateral $OMDN$ is concyclic, from which it follows that $\angle NDM + \angle NOM = 180°$ or $\angle BDN = \angle NOM$. Because $ABED$ is concyclic, we have $\angle BAE = \angle BDE$. Combining the above equations together, one has

$$\angle EOF = 2\angle NOM = 2\angle BDN = 2\angle BDE = 2\angle BAE.$$

Because BC is a diameter of Γ, it follows that

$$\angle EOF + \angle EAF = 2\angle BAE + 2\angle EAC = 2\angle BAC = 180°,$$

from which it follows that $AEOF$ is concyclic. Let ω denote the circumcircle of $AEOF$. Because O lies on the perpendicular bisector of the segment EF, O is the midpoint of the arc $\overset{\frown}{EF}$ (on ω), implying that AO bisects $\angle EAF$ and A, C, O are collinear.

3 Let n be a given positive integer. In the coordinate plane, consider the set of the points

$$\{P_1, P_2, \ldots, P_{4n+1}\}$$
$$= \{(x, y) \mid x \text{ and } y \text{ are integers with } xy = 0, \ |x| \leqslant n, \ |y| \leqslant n\}.$$

Determine the minimum of $(P_1P_2)^2 + (P_2P_3)^2 + \cdots + (P_{4n}P_{4n+1})^2 + (P_{4n+1}P_1)^2$. (Posed by Wang Xinmao)

Solution The answer is $16n - 8$.

Assume that $P_i = (x_i, y_i)$ for $1 \leqslant i \leqslant 4n + 1$. Set

$$P_{4n+2} = (x_{4n+2}, y_{4n+2}) = P_1 = (x_1, y_1).$$

We will show that the sum

$$S=\sum_{i=1}^{4n+1}(P_iP_{i+1})^2\geqslant 16n-8.$$

First, we show that this minimum can be obtained by setting

$(P_1, P_2, \ldots, P_{4n+1})=((2, 0), (4, 0), \ldots, (n, 0), (n-1, 0), \ldots, (1, 0), (0, 2), (0, 4), \ldots, (0, n), (0, n-1), \ldots, (0, 1), (-2, 0), (-4, 0), \ldots, (-n, 0), (-n+1, 0), \ldots, (-1, 0), (0, -2), (0, -4), \ldots, (0, -n), (0, -n+1), \ldots, (0, -1), (0, 0))$

for n even, and

$(P_1, P_2, \ldots, P_{4n+1})=((1, 0), (3, 0), \ldots, (n, 0), (n-1, 0), \ldots, (2, 0), (0, 1), (0, 3), \ldots, (0, n), (0, n-1), \ldots, (0, 2), (-1, 0), (-3, 0), \ldots, (-n, 0), (-n+1, 0), \ldots, (-2, 0), (0, -1), (0, -3), \ldots, (0, -n), (0, -n+1), \ldots, (0, -2), (0, 0))$

for n odd. This can be easily checked. For example, when $n=2m$ is even, our construction shows that

$$\begin{aligned}S&=4\left(\sum_{i=1}^{m-1}(P_iP_{i+1})^2+(P_mP_{m+1})^2+\sum_{i=m+1}^{2m-1}(P_iP_{i+1})^2\right)\\&\quad+3(P_{2m}P_{2m+1})^2+(P_{4n}P_{4n+1})^2+(P_{4n+1}P_1)^2\\&=4(4(m-1)+1+4(m-1))+3\times 5+1+4\\&=32m-8=16n-8.\end{aligned}$$

The exact same argument works when n is odd.

Note that

$$S=\sum_{i=1}^{4n+1}[(x_i-x_{i+1})^2+(y_i-y_{i+1})^2].$$

By symmetry, it suffices to show that for

$$\{x_1, x_2, \ldots, x_{4n+1}\} = \{\pm 1, \pm 2, \ldots, \pm n, \underbrace{0, \ldots, 0}_{2n+1\ 0s}\},$$

and we have

$$T_{n(x_1, x_2, \ldots, x_{4n+1})} = \sum_{i=1}^{4n+1} (x_i - x_{i+1})^2 \leqslant 8n - 4.$$

Indeed, for every positive integer n, we are going to prove that for $1 \leqslant m \leqslant n$ and

$$\{x_1, x_2, \ldots, x_{2m+2n+1}\} = \{\pm 1, \pm 2, \ldots, \pm m, \underbrace{0, \ldots, 0}_{2n+1\ 0s}\},$$

we have

$$T_{m(x_1, x_2, \ldots, x_{2m+2n+1})} = \sum_{i=1}^{2m+2n+1} (x_i - x_{i+1})^2 \leqslant 8m - 4.$$

We take induction on m. For $m=1$ and

$$\{x_1, x_2, \ldots, x_{2n+3}\} = \{\pm 1, \underbrace{0, \ldots, 0}_{2n+1\ 0s}\},$$

clearly $T_{1(x_1, x_2, \ldots, x_{2n+3})} = 8 \cdot 1 - 4 = 4$, while equality holds for

$$\{x_1, x_2, \ldots, x_{2n+3}\} = \{0, 1, 0, -1, 0, \underbrace{0, \ldots, 0}_{2n-2\ 0s}\}.$$

We assume the result is true for $m = k < n$, and we consider $m = k + 1$. By cyclic symmetry, we may assume that either

(1) $\{x_1, x_2, \ldots, x_{2m+2n+3}\} = (s, n, t, \ldots, u, -n, v, \ldots)$

or

(2) $\{x_1, x_2, \ldots, x_{2m+2n+3}\} = (s, n, -n, t, \ldots)$.

For (1), we note that

$$\begin{aligned} &T_{k+1, (s, n, t, \ldots, u, -n, v, \ldots)} \\ &= (n-s)^2 + (n-t)^2 - (s-t)^2 + (u+n)^2 - (u-v)^2 \\ &\quad + T_{k, (s, t, \ldots, u, v, \ldots)} \end{aligned}$$

$$= 2(n-s)(n-t) + 2(n+u)(n+v) + T_{k,(s,t,\ldots,u,v,\ldots)}$$
$$\geqslant 2[n-(n-1)][n-(n-2)] + 2[n+(-n+1)] \times$$
$$[n+(-n+2)] + 8k - 4$$
$$= 8(k+1) + 4,$$

by the induction hypothesis.

For (2), we note that

$$T_{k+1\,(s,n,-n,t,\ldots)}$$
$$= (n-s)^2 + (n+n)^2 + (n+t)^2 - (s-t)^2 + T_{k\,(s,t,\ldots)}$$
$$= 2n(n-s) + 2n(n+t) + 2(n^2+st) + T_{k\,(s,t,\ldots)}$$
$$\geqslant 2n + 2n + 2(2n-1) + 8k - 4 = 8(k+1) + 4,$$

by the induction hypothesis.

In every case, we have established our inductive process, completing our proof.

4 Let n be an integer greater than 3. The points $V_1, V_2, \ldots, V_n$, with no three collinear, lie on the plane. Some of the segments V_iV_j, with $1 \leqslant i < j \leqslant n$, are constructed. The points V_i and V_j are *neighbors* if V_iV_j is constructed. Initially, the chess pieces $C_1, C_2, \ldots, C_n$ are placed at the points $V_1, V_2, \ldots, V_n$ (not necessarily in that order), with exactly one piece at each point. In a move, one can choose some of the n chess pieces, and simultaneously relocate each of the chosen piece from its current position to one of its neighboring positions such that after the move, exactly one chess piece is at each point and no two chess pieces have exchanged their positions. A set of constructed segments is called *harmonic* if for any initial positions of the chess pieces each chess piece C_i $(1 \leqslant i \leqslant n)$ is at the point V_i after a finite number of moves.

Determine the minimum number of segments in a harmonic set. (Posed by Fu Yunhao)

Solution The answer is $n+1$.

For a harmonic set, we consider a graph G with $V_1, V_2, \ldots, V_n$ as its vertices and with the segments in the harmonic set as its edges.

First, we show that there are at least n edges in G. Note that G must be connected. Also note that each vertex must have degree at least 2, because when a chess piece is moved from V_i to V_j there is another piece moved from V_k (with $k \neq j$) to V_i. Hence, the total degree is at least $2n$, from which it follows that there are at least $2n = 2 = n$ edges.

Second, we show that there are at least $n+1$ edges. Assume that there are only n edges. In this connected graph, each vertex has exactly degree 2, and hence it must be a complete cycle. Without loss of generality, we may assume that the cycle $V_1 \to V_2 \to \cdots V_n \to V_1$ consists of all the edges. In this case, if C_1 and C_2 are placed at V_2 and V_1 initially, we cannot put them back to V_1 and V_2 simultaneously. This is because we can only rotate all the pieces along the cycle and cannot change their relative positions along the cycle.

Third, we show that $n+1$ edges is enough. We consider the graph G with the cycle $C_1: V_1 \to V_2 \to \cdots \to V_n \to V_1$ and one additional edge, V_2V_n. (This graph G now has the second cycle $C_2: V_2 \to V_3 \to \cdots \to V_n \to V_2$.) With this additional edge, we can switch the relative positions of the chess pieces along the cycle C_1. Indeed, without loss of generality, we may assume that C_i is at V_1 and C_j is at V_2 initially. Applying rotations on the cycle C_2, we can place C_j at V_n, i.e. the relative positions of C_i and C_j, along C_1, are switched. Because we can switch the positions of any two neighboring pieces in a finite amount of moves, we can place $C_1, C_2, \ldots, C_n$ in that order on the cycle C_1. We can then move each C_i to V_i by applying rotations along the cycle C_1.

Remark What if all the chess pieces have to be relocated at each move?

Second Day

(0800 - 1200; August 14, 2009)

5 let x, y, z be real numbers greater than or equal to 1. Prove that

$$(x^2 - 2x + 2)(y^2 - 2y + 2)(z^2 - 2z + 2) \leqslant (xyz)^2 - 2xyz + 2.$$

(Posed by Xiong Bin)

Solution Set $a = x - 1$, $b = y - 1$, $c = z - 1$. Then a, b, c are positive real numbers. Completing the square for trinomials of the form $A^2 - 2A + 2 = (A - 1)^2 + 1$ transforms the desired inequality into

$$\begin{aligned}&(a^2 + 1)(b^2 + 1)(c^2 + 1)\\ \leqslant\ &[(a + 1)(b + 1)(c + 1) - 1]^2 + 1\\ =\ &(abc + ab + bc + ca + a + b + c)^2 + 1.\end{aligned} \quad ①$$

From here, we can prove the inequality by direct expansions on both sides. Instead, we proceed in a slightly different way. By very simple expansions, we can see that

$$\begin{aligned}&(A^2 + 1)(B^2 + 1)\\ \leqslant\ &[(A + 1)(B + 1) - 1]^2 + 1\\ =\ &(AB + A + B)^2 + 1,\end{aligned} \quad ②$$

because the right-hand side has positive extra summands $2AB$, $2A^2B$, $2AB^2$. Applying ② twice [(first with $(A, B)=(a, b)$ and then with $(A, B)=(ab+a+b, c)$] yields ①.

6 The circle Γ_1, with radius r, is internally tangent to the circle Γ_2 at S. The chord AB of Γ_2 is tangent to Γ_2 at C. Let M be the midpoint of the arc $\overset{\frown}{AB}$ (not containing S), and let N be the foot of the perpendicular from M to the line AB. Prove that $AC \times CB = 2r \times MN$. (Posed by Ye Zhonghao)

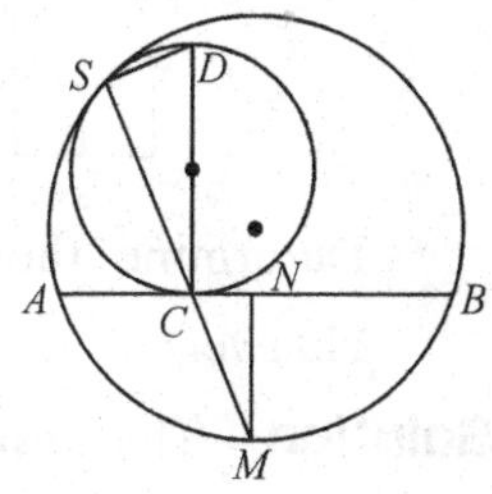

Solution It is well known that S, C, M are collinear. Indeed, consider the dilation centered at S that sends Γ_1 to Γ_2. Then the line AB is sent to the line l parallel to AB and tangent to Γ_2, i.e. the line tangent to Γ_2 at M (the midpoint of $\overset{\frown}{AB}$). Thus, this dilation sends C (the points of tangency of line the AB and Γ_1) to M (the points of tangency of the line l and Γ_2), from which it follows that S, C, M are collinear.

By the power-of-point theorem, we have $AC \times CB = SC \times CM$. It suffices to show that

$$SC \times CM = 2r \times MN \quad \text{or} \quad \frac{SC}{2r} = \frac{MN}{CM}. \qquad ①$$

Set $\angle MCN = \alpha$. Then $\angle SCA = \alpha$. By the extended sine law, we have $\frac{SC}{2r} = \sin\alpha$. In the right triangle MNC, we also have $\sin\alpha = \frac{MN}{CM}$. Combining the last two equations, we obtain ①.

(We can also derive ① by observing that the triangles MNC and CDS are similar.)

7 On a 10×10 chessboard, some $4n$ unit square fields are chosen to form a region R. This region R can be tiled by n 2×2 squares. If R can also be tiled by a combination of n pieces of the following types of shapes (with rotations allowed).

Determine the minimum value of n. (Posed by Zhu Huawei)

Solution The answer is $n = 4$. We call those two kinds of nonsquare tiles *ducks*.

First, the left-hand-side figure and the middle figure below show that $n = 4$ works.

	2	2		
	2	2	3	3
1	1		3	3
1	1	4	4	
		4	4	

	b	b		
	a	b	b	c
a	a		c	c
a	d	d	c	
		d	d	

×	×			×	×		
		×	×			×	×
		×	×			×	×
×	×			×	×		
×	×			×	×		
		×	×			×	×
		×	×			×	×
×	×			×	×		

Second, we show that n must be even. We mark the (infinite) chessboard with $\times$ in the pattern indicated in the right-hand-side figure. It is easy to see that each 2×2 covers exactly an even number of crosses (either two or four crosses) and each duck covers exactly an odd number of crosses (either one or three crosses). It follows that we must have an even number of ducks in R, i.e. n is even.

Third, we show that $n \geqslant 2$. If $n = 2$, then R can be tiled by two 2×2 squares. It is clear that these two squares much share a common edge. Hence, we can have only two possibilities:

It is not difficult to see that is not possible to tile either of the above configurations by a combinations of two ducks.

8 For positive integer n, $a_n = n\sqrt{5} - [n\sqrt{5}]$. Compute the maximum value and the minimum value of $a_1, a_2, \ldots, a_{2009}$. (For real number x, $[x]$ denotes the greatest integer less than or equal to x.) (Posed by Wang Zhixiong)

Solution Let $b_0 = 0$, $b_1 = 1$, $b_n = 4b_{n-2} + b_{n-1}$ $(n \geqslant 2)$. Then

$$b_n = \frac{(2+\sqrt{5})^n - (2-\sqrt{5})^n}{2\sqrt{5}}.$$

In particular, $b_6 = 1292$, $b_7 = 5473$.

For every $k = 1, 2, \ldots, 5473$, there are unique integers x_k, y_k such that $1292k = x_k + 5473y_k$, and $1 \leqslant x_k \leqslant 5473$. Since $(1292, 5473) = 1$, $x_1, x_2, \ldots, x_{5473}$ is a permutation of $1, 2, \ldots, 5473$, and it is clear that $\{y_k\}$ is nondecreasing:

$$y_1 \leqslant y_2 \leqslant \cdots \leqslant y_{5473} = 1291.$$

For our convenience, let $f(x) = x - [x]$. We have

$$\begin{aligned} f(x_k\sqrt{5}) &= f(1292k\sqrt{5} - 5473y_k\sqrt{5}) \\ &= f\left(\frac{(2+\sqrt{5})^6 - (2-\sqrt{5})^6}{2}k - \frac{(2+\sqrt{5})^7 - (2-\sqrt{5})^7}{2}y_k\right) \\ &= f(-(2-\sqrt{5})^6 k + (2-\sqrt{5})^7 y_k). \end{aligned}$$

Since

$$\begin{aligned} 0 &< (2-\sqrt{5})^6 k - (2-\sqrt{5})^7 y_k \\ &\leqslant 5473(2-\sqrt{5})^6 - 1291(2-\sqrt{5})^7 < 1, \end{aligned}$$

it follows that

$$f(x_k\sqrt{5}) = 1 - (2 - \sqrt{5})^6 k + (2 - \sqrt{5})^7 y_k,$$

and this is strictly decreasing.

Now, since $x_1 = 1292$, $x_{5473} = 5473$, $x_{5472} = 4181$, $x_{5471} = 2889$, $x_{5470} = 1597$, we see that a_{1292} attains the maximum and a_{1597} the minimum.

China Western Mathematical Olympiad

2008 (Guiyang, Guizhou)

The 8th China Western Mathematical Olympiad was held from October 30 to November 4, 2008 in Guiyang, Guizhou, China. The event was hosted by Guizhou Mathematical Society and HS affiliated to Guizhou Normal University.

The competition committee comprised Zhu Huawei, Wu Jianping, Chen Yonggao, Li Shenghong, Liu Shixiong, Feng Zhigang, Tang Lihua, Bian Hongping and Shi Xiaokang.

First Day

(0800 - 1200; November 1, 2008)

1 A sequence of real numbers $\{a_n\}$ is defined by $a_0 \neq 0, 1$, $a_1 = 1 - a_0$, $a_{n+1} = 1 - a_n(1 - a_n)$, $n = 1, 2, \ldots$. Prove that for any positive integer n, we have

$$a_0 a_1 \cdots a_n \left(\frac{1}{a_0} + \frac{1}{a_1} + \cdots + \frac{1}{a_n}\right) = 1.$$

(Posed by Li Shenghong)

Proof From the given condition, we have

$$\begin{aligned} 1 - a_{n+1} &= a_n(1 - a_n) = a_n a_{n-1}(1 - a_{n-1}) = \cdots \\ &= a_n \cdots a_1 (1 - a_1) = a_n \cdots a_1 a_0, \end{aligned}$$

i.e. $a_{n+1} = 1 - a_0 a_1 \cdots a_n$, $n = 1, 2, \ldots$.

By mathematical induction, when $n = 1$ the proposition holds. Assuming that it holds for $n = k$, then when $n = k + 1$ we have

$$\begin{aligned} & a_0 a_1 \cdots a_{k+1} \left(\frac{1}{a_0} + \frac{1}{a_1} + \cdots + \frac{1}{a_k} + \frac{1}{a_{k+1}}\right) \\ = {} & a_0 a_1 \cdots a_k \left(\frac{1}{a_0} + \frac{1}{a_1} + \cdots + \frac{1}{a_k}\right) a_{k+1} + a_0 a_1 \cdots a_k \\ = {} & a_{k+1} + a_0 a_1 \cdots a_k \\ = {} & 1. \end{aligned}$$

So it also holds when $n = k + 1$. Hence, it holds for any positive integer n.

2 In $\triangle ABC$, $AB = AC$, the inscribed circle I touches BC, CA, AB at points D, E and F respectively. P is a point on arc $\widehat{EF}$ (not containing D). Line BP intersects the circle I at another point Q, and lines EP, EQ meet line BC

at M, N respectively. Prove that

(1) P, F, B, M are concyclic;

(2) $\frac{EM}{EN} = \frac{BD}{BP}$.

(Posed by Bian Hongping)

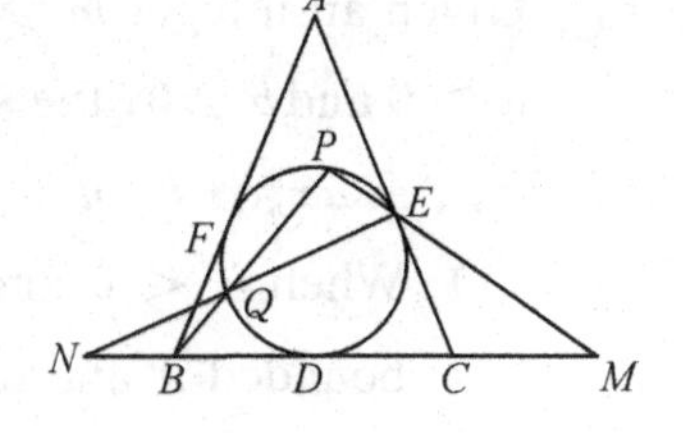

Proof (1) From the given condition, $EF \parallel BC$, so

$$\begin{aligned}\angle ABC &= \angle AFE = \angle AFP + \angle PFE \\ &= \angle PEF + \angle PFE = 180^\circ - \angle FPE,\end{aligned}$$

and thus P, F, B, M are concyclic.

(2) By the sine law, $EF \parallel BC$ and the fact that P, F, B, M are concyclic, we have

$$\frac{EM}{EN} = \frac{\sin\angle ENM}{\sin\angle EMN} = \frac{\sin\angle FEN}{\sin(\pi - \angle PFB)} = \frac{\sin\angle FPB}{\sin\angle PFB} = \frac{BF}{BP}.$$

Together with $BF = BD$, the proposition is proven.

3 Given an integer $m \geqslant 2$, and m positive integers a_1, a_2, ..., a_m, prove that there exist infinitely many positive integers n, such that $a_1 \cdot 1^n + a_2 \cdot 2^n + \cdots + a_m \cdot m^n$ is composite. (Posed by Chen Yonggao)

Proof Let p be a prime factor of $a_1 + 2a_2 + \cdots + ma_m$. By Fermat's little theorem, for any k and m satisfying $1 \leqslant k \leqslant m$, we have $k^p \equiv k \pmod{p}$. Thus, for any positive integer n, we have

$$\begin{aligned}a_1 \cdot 1^{p^n} + a_2 \cdot 2^{p^n} + \cdots + a_m \cdot m^{p^n} &\equiv a_1 + 2a_2 + \cdots + ma_m \\ &\equiv 0 \pmod{p}.\end{aligned}$$

Hence, $a_1 \cdot 1^{p^n} + a_2 \cdot 2^{p^n} + \cdots + a_m \cdot m^{p^n}$ $(n = 1, 2, \ldots)$ is composite.

4 Given an integer $m \geqslant 2$, and two real numbers a, b with $a > 0$ and $b \neq 0$, the sequence $\{x_n\}$ is such that $x_1 = b$ and $x_{n+1} = ax_n^m + b$, $n = 1, 2, \ldots$. Prove that:

(1) When $b < 0$ and m is even, the sequence $\{x_n\}$ is bounded if and only if $ab^{m-1} \geqslant -2$;

(2) When $b < 0$ and m is odd, or when $b > 0$, the sequence $\{x_n\}$ is bounded if and only if $ab^{m-1} \leqslant \dfrac{(m-1)^{m-1}}{m^m}$.

(Posed by Zhu Huawei and Fu Yunhao)

Proof (1) When $b < 0$ and m is even, in order that $ab^{m-1} < -2$, we should first have $ab^m + b > -b > 0$, and therefore $a(ab^m + b)^m + b > ab^m + b > 0$, i.e. $x_3 > x_2 > 0$. Using the fact that $ax^m + b$ is monotonically increasing on $(0, +\infty)$, it can be established that each succeeding term of the sequence $\{x_n\}$ is greater than its preceding term, and is greater than $-b$ starting from the second term.

Considering any three consecutive terms of the sequence x_n, x_{n+1}, x_{n+2}, $n = 2, 3, \ldots$, we have

$$\begin{aligned} x_{n+2} - x_{n+1} &= a(x_{n+1}^m - x_n^m) \\ &= a(x_{n+1} - x_n)(x_{n+1}^{m-1} + x_{n+1}^{m-2}x_n + \cdots + x_n^{m-1}) \\ &> amx_n^{m-1}(x_{n+1} - x_n) \\ &> am(-b)^{m-1}(x_{n+1} - x_n) \\ &> 2m(x_{n+1} - x_n) \\ &> x_{n+1} - x_n. \end{aligned}$$

It is obvious that the difference of any two consecutive terms of the sequence $\{x_n\}$ is increasing, and hence it is not bounded.

When $ab^{m-1} \geqslant -2$, mathematical induction is used to prove that each term of the sequence $\{x_n\}$ falls on the interval $[b, -b]$.

The first term b falls on the interval $[b, -b]$. Suppose that the term x_n satisfies the condition $b \leqslant x_n \leqslant -b$ for a particular n. Then $0 \leqslant x_n^m \leqslant b^m$, and hence

$$b = a \times 0^m + b \leqslant x_{n+1} \leqslant ab^m + b \leqslant -b.$$

Thus, the sequence $\{x_n\}$ is bounded if and only if $ab^{m-1} \geqslant -2$.

(2) When $b > 0$, each term of the sequence $\{x_n\}$ is positive. So, we first prove that $\{x_n\}$ is bounded if and only if the equation $ax^m + b = x$ has positive real roots.

Suppose that $ax^m + b = x$ has no positive real roots. In such a case, the minimum value of the function $p(x) = ax^m + b - x$ on the interval $(0, +\infty)$ is greater than zero. Let t be the minimum value. It follows that for any two consecutive terms of the sequence x_n and x_{n+1}, we have $x_{n+1} - x_n = ax_n^m - x_n + b$. Thus, each succeeding term of the sequence $\{x_n\}$ is greater than the preceding term by at least t. Hence, it is not bounded.

If the equation $ax^m + b = x$ has positive real roots, let x_0 be one of the positive real roots. Then, by using mathematical induction, we prove that each term of the sequence $\{x_n\}$ is less than x_0. Firstly, the first term b is less than x_0. Suppose that $x_n < x_0$ for a particular n. By virtue of the fact that $ax^m + b$ is increasing on the interval $[0, +\infty)$, it can be established that

$$x_{n+1} = ax_n^m + b < ax_0^m + b = x_0.$$

Therefore, the sequence is bounded.

Further, the equation $ax^m + b = x$ has positive roots if and only if the minimum value of $ax^{m-1} + \dfrac{b}{x}$ on the interval $(0, +\infty)$ is not greater than 1, whereas the minimum value of $ax^{m-1} + \dfrac{b}{x}$ can be determined by mean inequality, i.e.

$$ax^{m-1}+\frac{b}{x}=ax^{m-1}+\frac{b}{(m-1)x}+\cdots+\frac{b}{(m-1)x}\geqslant m\sqrt[m]{\frac{ab^{m-1}}{(m-1)^{m-1}}}.$$

As such, the sequence $\{x_n\}$ is bounded if and only if $m\sqrt[m]{\frac{ab^{m-1}}{(m-1)^{m-1}}}\leqslant 1$, i.e. $ab^{m-1}\leqslant\frac{(m-1)^{m-1}}{m^m}$.

When $b<0$, and m is odd, let $y_n=-x_n$. Then $y_1=-b>0$, $y_{n+1}=ay_n^m+(-b)$, showing that the sequence $\{x_n\}$ is bounded if and only if the sequence $\{y_n\}$ is bounded. Thus, by using the above reasoning , it can be proven that (2) holds.

Second Day

(0800 - 1200; November 2, 2008)

5 Four frogs are positioned at four points on a straight line such that the distance between any two neighboring points is one unit of length. Suppose that every frog can jump to its corresponding point of reflection, by taking any one of the other three frogs as the reference point. Prove that there is no case where the distances between any two neighboring points, where the frogs stay, we all equal to 2008 unit of length. (Posed by Liu Shixiong)

Proof Without loss of generality, we may think of the initial positioning of the four frogs as being on the real number line at points 1, 2, 3, and 4. Further, it can be established that the frogs at odd number positions remain at odd number positions after each jump, and likewise for frogs at even number positions. Thus, no matter after how many jumps, there are two frogs remaining at odd number positions while the other two frogs remain at even number positions. Therefore, in order that the distances between any two neighboring frogs are all equal to 2008, all the frogs need to stay at points which are either all odd or all even, which is contrary to the actual situation. Hence, the proposition is proven.

6 Given $x, y, z \in (0, 1)$ satisfying

$$\sqrt{\frac{1-x}{yz}}+\sqrt{\frac{1-y}{zx}}+\sqrt{\frac{1-z}{xy}}=2,$$

find the maximum value of xyz. (Posed by Tang Lihua)

Solution Denote $u=\sqrt[6]{xyz}$. Then by the given condition and mean inequality,

$$\begin{aligned}2u^3 &= 2\sqrt{xyz}=\frac{1}{\sqrt{3}}\sum\sqrt{x(3-3x)}\\ &\leqslant\frac{1}{\sqrt{3}}\sum\frac{x+(3-3x)}{2}=\frac{3\sqrt{3}}{2}-\frac{1}{\sqrt{3}}(x+y+z)\\ &\leqslant\frac{3\sqrt{3}}{2}-\sqrt{3}\times\sqrt[3]{xyz}=\frac{3\sqrt{3}}{2}-\sqrt{3}u^2.\end{aligned}$$

Therefore, $4u^3+2\sqrt{3}u^2-3\sqrt{3}\leqslant 0$, i.e.

$$(2u-\sqrt{3})(2u^2+2\sqrt{3}u+3)\leqslant 0,$$

and thus $u\leqslant\frac{\sqrt{3}}{2}$. Following this, we have $xyz\leqslant\frac{27}{64}$, and equality holds when $x=y=z=\frac{3}{4}$. Hence, the maximum is $\frac{27}{64}$.

7 For a given positive integer n, find the greatest positive integer k, such that there exist three sets of k distinct nonnegative integers, $A=\{x_1, x_2, \ldots, x_k\}$, $B=\{y_1,$

$y_2, \ldots, y_k\}$ and $C = \{z_1, z_2, \ldots, z_k\}$ with $x_j + y_j + z_j = n$ for any $1 \leqslant j \leqslant k$. (Posed by Li Shenghong)

Solution By the given condition, we have

$$kn \geqslant \sum_{i=1}^{k}(x_i + y_i + z_i) \geqslant 3\sum_{i=0}^{k-1} i = \frac{3k(k-1)}{2},$$

and then $k \leqslant \left[\frac{2n}{3}\right] + 1$.

The following illustrates the case of $k = \left[\frac{2n}{3}\right] + 1$:

Set $m \in \mathbf{Z}^+$. When $n = 3m$, for $1 \leqslant j \leqslant m+1$, let $x_j = j-1$, $y_j = m+j-1$, $z_j = 2m - 2j + 2$; for $m+2 \leqslant j \leqslant 2m+1$, let $x_j = j-1$, $y_j = j-m-2$, $z_j = 4m-2j+3$, and the result is obvious. When $n = 3m+1$, for $1 \leqslant j \leqslant m$, let $x_j = j-1$, $y_j = m+j$, $z_j = 2m-2j+2$; for $m+1 \leqslant j \leqslant 2m$, let $x_j = j+1$, $y_j = j-m-1$, $z_j = 4m+1-2j$; and $x_{2m+1} = m$, $y_{2m+1} = 2m+1$, $z_{2m+1} = 0$ will lead to the expected result. When $n = 3m+2$, for $1 \leqslant j \leqslant m+1$, let $x_j = j-1$, $y_j = m+j$, $z_j = 2m-2j+3$; for $m+2 \leqslant j \leqslant 2m+1$, let $x_j = j$, $y_j = j-m-2$, $z_j = 4m-2j+4$; and $x_{2m+2} = 2m+2$, $y_{2m+2} = m$, $z_{2m+2} = 0$, and the result follows.

In summary, the maximum value of k is $\left[\frac{2n}{3}\right] + 1$.

8 Let P be an interior point of a regular n-gon $A_1A_2\cdots A_n$; the lines A_iP meet the regular n-gon $A_1A_2\cdots A_n$ at another point B_i, where $i = 1, 2, \ldots, n$. Prove that

$$\sum_{i=1}^{n} PA_i \geqslant \sum_{i=1}^{n} PB_i.$$

(Posed by Feng Zhigang)

Proof Denote $t = \left[\frac{n}{2}\right]+1$, and let $A_{n+j} = A_j$, $j = 1, 2, \ldots, n$.

Noting that the distance between any vertex of a regular n-gon and a point on its side is not greater than its longest diagonal d, we therefore have, for any $1 \leqslant i \leqslant n$,

$$A_iP + PB_i = A_iB_i \leqslant d. \quad ①$$

Furthermore, using the fact that the sum of any two sides of a triangle is longer than the third side, we have, for any $1 \leqslant i \leqslant n$,

$$A_iP + PA_{i+t} \geqslant A_iA_{i+t} = d. \quad ②$$

Summing up ①, ② for $i = 1, 2, \ldots, n$, we have

$$\sum_{i=1}^{n}(A_iP + PA_{i+t}) \geqslant nd \geqslant \sum_{i=1}^{n}(A_iP + PB_i),$$

i.e.

$$2\sum_{i=1}^{n} PA_i \geqslant \sum_{i=1}^{n} A_iP + \sum_{i=1}^{n} PB_i,$$

following which the proposition is proven.

2009 (Kunming, Yunnan)

The 9th China Western Mathematical Olympiad was held from October 26 to November 1, 2009 in Kunming, Yunnan, China. The event was hosted by Yunnan Mathematical Society and HS affiliated to Yunnan Normal University.

The competition committee comprised Xiong Bin, Wu Jianping, Chen Yonggao, Li Shenghong, Ieng Tak Leong, Liu

Shixiong, Feng Zhigang, Li Jianping, Bian Hongping and Li Qiusheng.

First Day

(0800 - 1200; October 29, 2009)

1 Let M be a subset of $\mathbf{R}$ obtained by deleting finitely many real numbers from $\mathbf{R}$. Prove that for any given positive number n, there exists a polynomial $f(x)$ of degree n such that all its coefficients and its n real roots are in M. (Posed by Feng Zhigang)

Proof Let $T = \{|x| \in \mathbf{R} \mid x \notin M\}$, which is a finite set by definition, and $a = \max T$. Choose any real number $k > \max\{|a|, 1\}$; then $-k \notin T$ and hence $-k \in M$.

For any positive integer n, define $f(x) = k(x+k)^n$. Then it follows from $k > 1$ that $\deg f(x) = n$, and the coefficient of x^m in $f(x)$ is given by

$$k \times \binom{n}{m} \times k^{n-m} \geqslant k.$$

Hence, all the coefficients of $f(x)$ are not in T, and thus are in M. As the roots of $f(x)$ are all $-k$ with multiplicity n, they are in M. It follows that the polynomial $f(x) = k(x+k)^n$ satisfies the condition.

2 Let $n \geqslant 3$ be any given integer. Determine the smallest positive integer k, for which there exists a set A of k real numbers and n real numbers $x_1, x_2, \ldots, x_n$, which are distinct from each other such that

$$x_1 + x_2, x_2 + x_3, \ldots, x_{n-1} + x_n, x_n + x_1$$

are all in the set A. (Posed by Xiong Bin)

Solution Let $m_1 = x_1 + x_2$, $m_2 = x_2 + x_3$, $\ldots$, $m_{n-1} = x_{n-1} + x_n$, $m_n = x_n + x_1$.

First, note that $m_1 \neq m_2$, otherwise $x_1 = x_3$, which contradicts the fact that x_i are distinct. Similarly, $m_i \neq m_{i+1}$, for $i = 1, 2, \ldots, n$, where $m_{n+1} = m_1$, as usual. It follows that $k \geqslant 2$.

For $k = 2$, let $A = \{a, b\}$, where $a \neq b$. It follows that

$$(1)\begin{cases} x_1 + x_2 = a, \\ x_2 + x_3 = b, \\ \ldots \\ x_{n-1} + x_n = a, \\ x_n + x_1 = b, \end{cases} \qquad \text{(if } n \text{ is even)}$$

or

$$(2)\begin{cases} x_1 + x_2 = a, \\ x_2 + x_3 = b, \\ \ldots \\ x_{n-1} + x_n = b, \\ x_n + x_1 = a. \end{cases} \qquad \text{(if } n \text{ is odd)}$$

For (2), we have $x_n = x_2$, which is possible. For (1), it follows that

$$\begin{aligned} \frac{n}{2}a &= (x_1 + x_2) + (x_3 + x_4) + \cdots + (x_{n-1} + x_n) \\ &= (x_2 + x_3) + (x_4 + x_5) + \cdots + (x_n + x_1) \\ &= \frac{n}{2}b, \end{aligned}$$

and hence $a = b$, which is impossible again. It follows that $k \geqslant 3$.

For $k = 3$, one can construct a valid example as follows:

define $x_{2k-1} = k\,(k \geqslant 1)$ and $x_{2k} = n+1-k\ (k \geqslant 1)$. When n is even,

$$x_i + x_{i+1} = \begin{cases} n+1, & \text{if } i \text{ is odd,} \\ n+2, & \text{if } i \text{ is even and } i < n, \\ \dfrac{n}{2}+2, & \text{if } i = n, \text{ where } x_{n+1} = x_n. \end{cases}$$

When n is odd,

$$x_i + x_{i+1} = \begin{cases} n+1, & \text{if } i \text{ is odd and } i < n, \\ n+2, & \text{if } i \text{ is even,} \\ \dfrac{n-1}{2}+2, & \text{if } i = n, \text{ where } x_{n+1} = x_n. \end{cases}$$

Therefore, the smallest positive integer k is 3.

3 Let H be the orthocenter of an acute triangle ABC, and D be the midpoint of the side BC. A line passing through the point H meets the sides AB, AC at the points F, E respectively, such that $AE = AF$. The ray DH meets the circumcircle of $\triangle ABC$ at the point P.

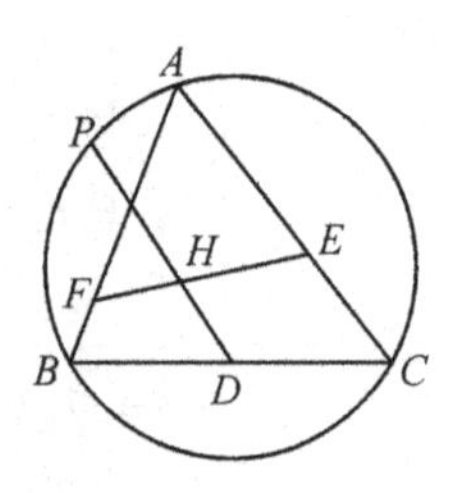

Prove that P, A, E, F are concyclic. (Posed by Bian Hongping)

Proof On the ray HD, mark a point M such that $HD = DM$. Join the segments BM, CM, BH and CH. As D is the midpoint of BC, the quadrilateral $BHCM$ is a parallelogram, and so

$$\angle BMC = \angle BHC = 180^\circ - \angle BAC.$$

Hence,

$$\angle BMC + \angle BAC = 180^\circ,$$

and the point M lies on the circumcircle of $\triangle ABC$. Join the segments PB, PC, PE and PF. It follows from $AE = AF$ that

$$\angle BFH = \angle CEH. \qquad ①$$

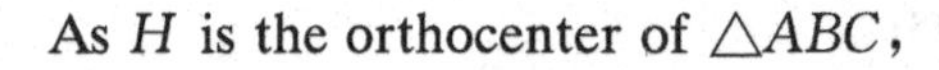

As H is the orthocenter of $\triangle ABC$,

$$\angle HBF = 90^\circ - \angle BAC = \angle HCE. \qquad ②$$

From① and②, one has $\triangle BFH \backsim \triangle CEH$, so

$$\frac{BF}{BH} = \frac{CE}{CH}.$$

As $BHCM$ is a parallelogram, $BH = CM$, $CH = BM$, and hence

$$\frac{BF}{CM} = \frac{CE}{BM}. \qquad ③$$

And D is the midpoint of BC. Thus, $S_{\triangle PBM} = S_{\triangle PCM}$, and so

$$\frac{1}{2}BP \times BM \times \sin\angle MBP = \frac{1}{2}CP \times CM \times \sin\angle MCP.$$

It follows from $\angle MBP + \angle MCP = 180^\circ$ that

$$BP \times BM = CP \times CM. \qquad ④$$

With ③ and ④, one has $\frac{BF}{BP} = \frac{CE}{CP}$. From $\angle PBF = \angle PCE$, one has $\triangle PBF \backsim \triangle PCE$, and hence $\angle PFB = \angle PEC$, and therefore $\angle PFA = \angle PFA$. Consequently P, A, E and F are concyclic.

4 Prove that for any given positive integer k, there exist infinitely many positive integers n, such that the numbers

$$2^n + 3^n - 1,\ 2^n + 3^n - 2,\ \ldots,\ 2^n + 3^n - k$$

are all composite. (Posed by Chen Yonggao)

Proof For any given positive integer k, choose positive integer m sufficiently large such that $2^m + 3^m - k > 1$. Consider the following k integers:

$$2^m + 3^m - 1,\ 2^m + 3^m - 2,\ \ldots,\ 2^m + 3^m - k,$$

all of which are larger than 1. From each of these integers, pick a prime factor: p_1, p_2, ..., p_k, and let

$$n_t = m + t(p_1 - 1)(p_2 - 1)\cdots(p_k - 1),$$

where t is an arbitrary positive integer. For any fixed integer i $(1 \leqslant i \leqslant k)$, one has $2^{n_t} \equiv 2^m \pmod{p_i}$. In fact, if $p_i = 2$, then the result is obvious. Assuming that $p_i \neq 2$, it follows from Fermat's little theorem that

$$2^{n_t} = 2^m \cdot 2^{t(p_1-1)(p_2-1)\ldots(p_k-1)} \equiv 2^m \cdot 1 = 2^m \pmod{p_i}.$$

Similarly, one has $3^{n_t} \equiv 3^m \pmod{p_i}$. Observe that

$$2^{n_t} + 3^{n_t} - i \equiv 2^m + 3^m - i \equiv 0 \pmod{p_i},$$

$$2^{n_t} + 3^{n_t} - i > 2^m + 3^m - i.$$

Hence, $2^{n_t} + 3^{n_t} - i$ is a composite number.

Therefore, n_t is one of the positive integers n such that

$$2^n + 3^n - 1,\ 2^n + 3^n - 2,\ \ldots,\ 2^n + 3^n - k$$

are all composite. As t is arbitrarily chosen, there are infinitely many such positive integers satisfying the conditions above.

Second Day

(0800 - 1200; October 30, 2009)

5 Let $\{x_n\}$ be a sequence such that $x_1 \in \{5, 7\}$, and $x_{n+1} \in \{5^{x_n}, 7^{x_n}\}$, for $n = 1, 2, \ldots$. Determine all the possible cases of the last two digits of x_{2009}. (Posed by Ieng Tak Leong)

Solution Let $n = 2009$. Then we have the following three cases:

(1) If $x_n = 7^{5^{x_{n-2}}}$, then $x_n \equiv 7 \pmod{100}$.

Since both 5 and 7 are odd, x_k $(1 \leqslant k \leqslant n)$ all are odd. $5^{x_{n-2}} \equiv 1 \pmod 4$, i.e. $5^{x_{n-2}} = 4k + 1$ for some positive integer k.

If $k, m \geqslant 0$, it follows from mathematical induction that

$$7^{4k+m} \equiv 7^m \pmod{100},$$

and hence

$$x_n = 7^{5^{x_{n-2}}} = 7^{4k+1} \equiv 7 \pmod{100}.$$

(2) If $x_n = 7^{7^{x_{n-2}}}$, then $x_n \equiv 43 \pmod{100}$.

$$7^{x_{n-2}} \equiv (-1)^{x_{n-2}} \equiv -1 \equiv 3 \pmod 4,$$

so $7^{x_{n-2}} = 4k + 3$, for some positive integer k. Hence,

$$x_n = 7^{7^{x_{n-2}}} = 7^{4k+3} \equiv 7^3 \equiv 43 \pmod{100}.$$

(3) If $x_n = 5^{x_{n-1}}$, then $x_n \equiv 25 \pmod{100}$.

It follows from mathematical induction that if $n \geqslant 2$, then $5^n \equiv 25 \pmod{100}$. It is obvious that $x_{n-1} > 2$, and hence $x_n = 5^{x_{n-1}} \equiv 25 \pmod{100}$.

Therefore, the possible cases of the last two digits of x_{2009} are 07, 25, 43.

6 Let D be a point on the side BC of an acute triangle ABC. The circle with diameter BD meets the lines AB and AD respectively at the points X and P, which are different from the points B and D. The circle with diameter CD meets the lines AC and AD respectively at the points Y and Q, which are different from the points C and D. Through the point A draw two lines which are perpendicular to PX and QY with the feet of perpendicular M and N respectively.

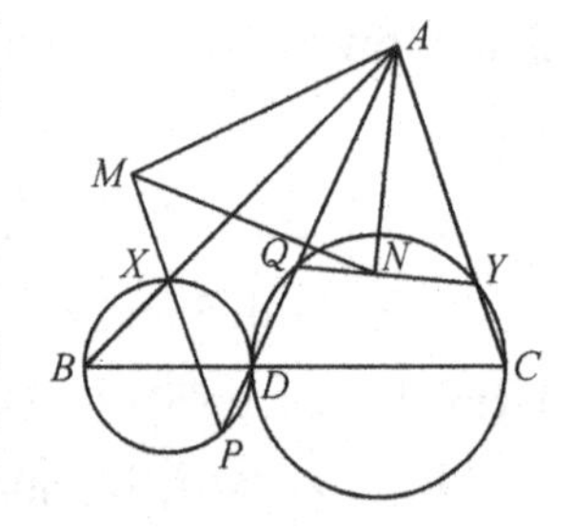

Prove that $\triangle AMN \backsim \triangle ABC$ if and only if the line AD passes through the circumcenter of $\triangle ABC$. (Posed by Li Qiusheng)

Proof Join the segments XY and DX. It follows from the given conditions that B, P, D, X are concyclic, and C, Y, Q, D are concyclic. Then

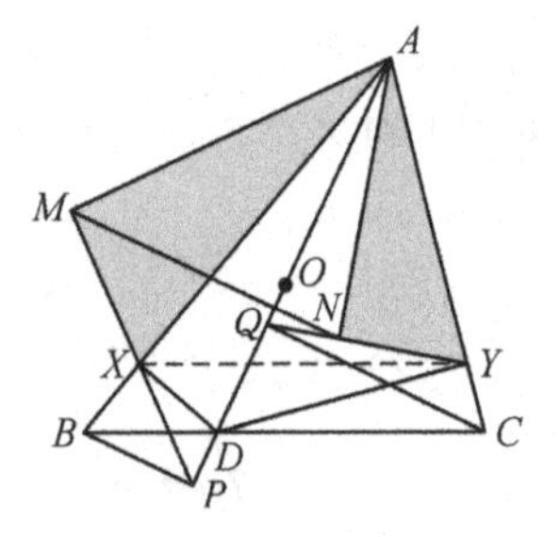

$$\begin{aligned}\angle AXM &= \angle BXP = \angle BDP \\ &= \angle QDC = \angle AYN.\end{aligned}$$

And it follows from $\angle AMX = \angle ANY = 90^\circ$ that $\triangle AMC \backsim \triangle ANY$, so $\angle MAX = \angle NAY$ and $\dfrac{AM}{AX} = \dfrac{AN}{AY}$, and hence $\angle MAN = \angle XAY$.

Combining the two results above, we have $\triangle AMN \backsim \triangle AXY$. So we get

$$\begin{aligned}&\triangle AMN \backsim \triangle ABC \Leftrightarrow \triangle AXY \backsim \triangle ABC \\ &\Leftrightarrow XY \parallel BC \Leftrightarrow \angle DXY = \angle XDB.\end{aligned}$$

As A, X, D, Y are concyclic, we have $\angle DXY = \angle DAY$. As $\angle XDB = 90^\circ - \angle ABC$,

$$\angle DXY = \angle XDB \Leftrightarrow \angle DAC = 90^\circ - \angle ABC,$$

which is equivalent to the fact that the line AD passes through the circumcenter of $\triangle ABC$.

Hence, $\triangle AMN \backsim \triangle ABC$ if and only if AD passes through the circumcenter of $\triangle ABC$.

7 There are n ($n > 12$) students participating in a mathematics contest. The examination paper consists of 15 fill-in-the-blank questions. For each question, the score of a correct answer is 1 point, and no point will be awarded if the answer is wrong or left blank. After analyzing all the possible cases of score distributions of these n students, one finds out that if the sum of total scores of any 12 students is not less than 36 points, then there are at least 3 students among these n students who answer at least 3 identical questions correctly. Determine the smallest possible value of n. (Posed by Liu Shixiong)

Solution The smallest n is 911. We divide the proof into two parts:

(1) We first prove that $n = 911$ satisfies the conditions. If each student answers at least 3 questions correctly, then for any student there are $\binom{15}{3} = 455$ ways for him to have exactly 3 correct answers. If there are 911 students participating in the contest, it follows from the pigeonhole principle that there are at least 3 students having 3 identical correct answers.

If there is a student X whose score is not more than 2, then the number of remaining students with a score not more than 3 cannot exceed 10; otherwise, pick any 11 of these students together with X, and the sum of their total scores is less than 36 points. Then there are more than $911 - 11 = 900$ students in the rest, such that each of them has a score less than 4. Since $\binom{4}{3} = 4$, and $4 \times 900 > 455 \times 2$, there are at least 3 students answering 3 identical questions correctly.

(2) There are 910 students participating in the contest. Divide them into $455 = \binom{15}{3}$ groups, and there are exactly 2 students in each group. In each group, both students have the identical answers with only 3 correct answers indexed by the group label.

8 Let $a_1, a_2, \ldots, a_n (n \geqslant 3)$ be real numbers satisfying $a_1 + a_2 + \cdots + a_n = 0$, and

$$2a_k \leqslant a_{k-1} + a_{k+1}, \text{ for } k = 2, 3, \ldots, n-1.$$

Determine the smallest $\lambda(n)$, such that for any $k \in \{1, 2, \ldots, n\}$, one has

$$|a_k| \leqslant \lambda(n) \cdot \max\{|a_1|, |a_n|\}.$$

(Posed by Li Shenghong)

Solution $\lambda(n)_{\min} = \frac{n+1}{n-1}$. First, define a sequence $\{a_k\}$ as follows: $a_1 = 1$, $a_2 = -\frac{n+1}{n-1}$ and

$$a_k = -\frac{n+1}{n-1} + \frac{2n(k-2)}{(n-1)(n-2)}, \text{ for } k = 3, 4, \ldots, n.$$

Then $a_1 + a_2 + \cdots + a_n = 0$, and $2a_k \leqslant a_{k-1} + a_{k+1}$ for $k = 2, 3, \ldots,$

$n-1$. In this case, it is easy to check that $\lambda(n) \geqslant \frac{n+1}{n-1}$.

Next, suppose that $\{a_k\}$ is a sequence satisfying the two conditions stated in the questions. We will prove that the following inequalities hold:

$$a_k \leqslant \frac{n+1}{n-1}\max\{|a_1|, |a_n|\}, \text{ for all } k \in \{1, 2, \ldots, n\}.$$

As $2a_k \leqslant a_{k-1} + a_{k+1}$, so $a_{k+1} - a_k \geqslant a_k - a_{k-1}$,

$$a_n - a_{n-1} \geqslant a_{n-1} - a_{n-2} \geqslant \cdots \geqslant a_2 - a_1.$$

Then, using the telescope sum, we have

$$\begin{aligned}
&(k-1)(a_n - a_1)\\
&= (k-1)[(a_n - a_{n-1}) + (a_{n-1} - a_{n-2}) + \cdots + (a_2 - a_1)]\\
&\geqslant (n-1)[(a_k - a_{k-1}) + (a_{k-1} - a_{k-2}) + \cdots + (a_2 - a_1)]\\
&= (n-1)(a_k - a_1),
\end{aligned}$$

and hence

$$\begin{aligned}
a_k &\leqslant \frac{k-1}{n-1}(a_n - a_1) + a_1\\
&= \frac{1}{n-1}[(k-1)a_n + (n-k)a_1]. \qquad ①
\end{aligned}$$

Similarly, for any fixed k, where $k \notin \{1, n\}$, and for any $j \in \{1, 2, \ldots, k\}$, we have

$$a_j \leqslant \frac{1}{k-1}[(j-1)a_k + (k-j)a_1].$$

Hence, for $j \in \{k, k+1, \ldots, n\}$, we get

$$a_j \leqslant \frac{1}{n-1}[(j-k)a_n + (n-j)a_k].$$

Consequently, we have

$$\sum_{j=1}^{k} a_j \leqslant \frac{1}{k-1} \sum_{j=1}^{k} [(j-1)a_k + (k-j)a_1]$$
$$= \frac{k}{2}(a_1 + a_k),$$

$$\sum_{j=k}^{n} a_j \leqslant \frac{1}{n-k} \sum_{j=k}^{n} [(j-k)a_n + (n-j)a_k]$$
$$= \frac{n+1-k}{2}(a_k + a_n).$$

Summing up these two inequalities, we find that

$$a_k = \sum_{j=1}^{k} a_j + \sum_{j=k}^{n} a_j \leqslant \frac{k}{2}(a_1 + a_k) + \frac{n+1-k}{2}(a_k + a_n)$$
$$= \frac{k}{2}a_1 + \frac{n+1}{2}a_k + \frac{n+1-k}{2}a_n.$$

Therefore, we have

$$a_k \geqslant -\frac{1}{n-1}[ka_1 + (n+1-k)a_n]. \qquad ②$$

From ① and ②, for $k = 2, 3, \ldots, n-1$, we obtain

$$|a_k| \leqslant \max\left\{\frac{1}{n-1}|(k-1)a_n + (n-k)a_1|,\ \frac{1}{n-1}|ka_1 + (n+1-k)a_n|\right\}$$
$$\leqslant \frac{n+1}{n-1}\max\{|a_1|, |a_n|\}.$$

Consequently, it follows that $\lambda(n)_{\min} = \dfrac{n+1}{n-1}$.

China Southeastern Mathematical Olympiad

2009 (Nanchang, Jiangxi)

First Day

(0800 - 1200; July 28, 2007)

1 Find the integer solutions of the function $x^2 - 2xy + 126y^2 = 2009$. (Posed by Zhang Pengcheng)

Solution Suppose that the integers x, y satisfy

$$x^2 - 2xy + 126y^2 - 2009 = 0.$$

Looking at this as a quadratic function of x,

$$\Delta = 4y^2 - 4 \times (126y^2 - 2009) = 500(4^2 - y^2) + 36$$

should be a square number.

If $y^2 > 4^2$, then $\Delta < 0$. So $y^2 < 4^2$, when $y^2 \in \{0, 1^2, 2^2, 3^2\}$, $\Delta \in \{8036, 7536, 6036, 3536\}$ is not a square number.

For $y^2 = 4^2$, $\Delta = 500(4^2 - y^2) + 36 = 6^2$. According to $y = 4$, one can get $x = 1$ or 7; according to $y = -4$, one can get $x = -1$ or -7.

All the integer solutions are $(x, y) = (1, 4), (7, 4), (-1, -4), (-7, -4)$.

2 For a convex pentagon $ABCDE$, $AB = DE$, $BC = EA$, $AB \neq EA$, and B, C, D, E are concyclic. Prove that A, B, C, D are concyclic if and only if $AC = AD$. (Posed by Xiong Bin)

Solution First, if A, B, C, D are concyclic, by $AB = DE$ and $BC = EA$ we have $\angle BAC = \angle EDA$, $\angle ACB = \angle DAE$, so $\angle ABC = \angle DEA$, which means that $AC = AD$.

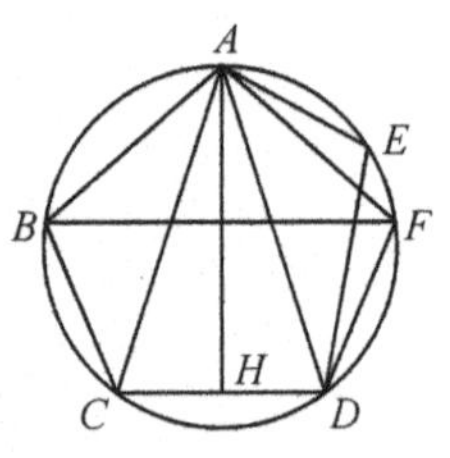

Second, if $AC = AD$, let O be the center of the circle (B, C, D, E are on the circle). Then O is on the perpendicular bisector (AH) of CD. Let F is the symmetric point of B by the line AH. Then F is on the circle O. $AB \neq EA$, so $DE \neq DF$, and thus E, F are not the same points. Now one can see that $\triangle AFD \cong \triangle ABC$, with $AB = DE$, $BC = EA$, and one can get $\triangle AED \cong \triangle CBA$, and so $\triangle AED \cong \triangle DFA$, which indicates $\angle AED = \angle DFA$, and therefore A, E, F, D is concyclic. This means that A is on the circle O, and A, B, C, D are concyclic.

3 Let x, y, z be positive numbers, and $\sqrt{a} = x(y-z)^2$, $\sqrt{b} = y(z-x)^2$, $\sqrt{c} = z(x-y)^2$. Prove that $a^2 + b^2 + c^2 \geqslant 2(ab + bc + ca)$. (Posed by Tang Lihua)

Solution

$$\sqrt{b} + \sqrt{c} - \sqrt{a} = -(y+z)(z-x)(x-y),$$
$$\sqrt{c} + \sqrt{a} - \sqrt{b} = -(z+x)(x-y)(y-z),$$
$$\sqrt{a} + \sqrt{b} - \sqrt{c} = -(x+y)(y-z)(z-x),$$

so

$$\begin{aligned}&(\sqrt{b}+\sqrt{c}-\sqrt{a})(\sqrt{c}+\sqrt{a}-\sqrt{b})(\sqrt{a}+\sqrt{b}-\sqrt{c})\\ =&-(y+z)(z+x)(x+y)[(y-z)(z-x)(x-y)]^2\\ \leqslant&\ 0.\end{aligned}$$

We can get

$$\begin{aligned}&2(ab+bc+ca)-(a^2+b^2+c^2)\\ =&(\sqrt{a}+\sqrt{b}+\sqrt{c})(\sqrt{b}+\sqrt{c}-\sqrt{a})(\sqrt{c}+\sqrt{a}-\sqrt{b})(\sqrt{a}+\sqrt{b}-\sqrt{c})\\ \leqslant&\ 0.\end{aligned}$$

This means that $a^2 + b^2 + c^2 \geqslant 2(ab + bc + ca)$.

4 There are given 12 red points on a circle. Find the minimum of n, such that there exist n triangles, whose vertices are red points, satisfying every chord with red endpoints being a side of one triangle. (Posed by Tao Pingsheng)

Solution Let the set of 12 red points be $A = \{A_1, A_2, \ldots, A_{12}\}$. From A_1, one can get 11 chords with red points, but every triangle with vertex A_1 has two such chords, and so the 11 chords should be in at least 6 triangles with A_1 being an endpoint. The same applies to A_i $(i = 2, 3, \ldots, 12)$; we need

$12 \times 6 = 72$ triangles, and every triangle has 3 points. So $n \geqslant \frac{72}{3} = 24$.

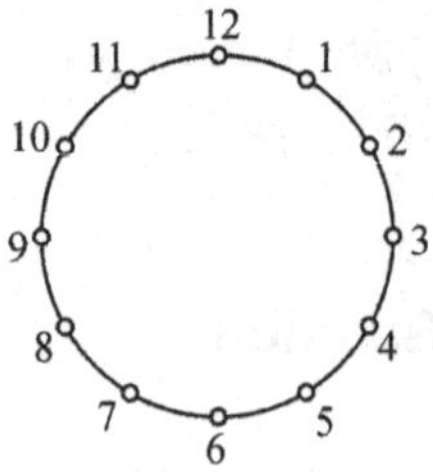

On the other hand, the following example shows that n can be 24.

Consider a circle whose perimeter is 12, and the 12 red points are on the circle with equal distance. The number of chords with red endpoint is $\binom{12}{2} = 66$. If the length of the minor arc to a chord is k, then say "the chord belongs to k." So we have only six kinds of chords, and the number of chords belonging to 1, 2, ..., 5 is 12, and that belonging to 6 is 6

If three chords belonging to a, b, c $(a \leqslant b \leqslant c)$ can be the sides of a triangle, then $a + b = c$ or $a + b + c = 12$. So the triangles $(a, b, c) \in \{(1, 1, 2), (2, 2, 4), (3, 3, 6), (2, 5, 5), (1, 2, 3), (1, 3, 4), (1, 4, 5), (1, 5, 6), (2, 3, 5), (2, 4, 6), (3, 4, 5), (4, 4, 4)\}$.

Now we give an example:

The number of (1, 2, 3) triangles is 6, whose vertices are

$$\{2, 3, 5\}, \{4, 5, 7\}, \{6, 7, 9\},$$
$$\{8, 9, 11\}, \{10, 11, 1\}, \{12, 1, 3\}.$$

The number of (1, 5, 6) triangles is 6, whose vertices are

$$\{1, 2, 7\}, \{3, 4, 9\}, \{5, 6, 11\},$$
$$\{7, 8, 1\}, \{9, 10, 3\}, \{11, 12, 5\}.$$

The number of (2, 3, 5) triangles is 6, whose vertices are

$$\{2, 4, 11\}, \{4, 6, 1\}, \{6, 8, 3\},$$
$$\{8, 10, 5\}, \{10, 12, 7\}, \{12, 2, 9\}.$$

The number of (4, 4, 4) triangles is 3, whose vertices are {1, 5, 9}, {2, 6, 10}, {3, 7, 11}.

The number of (2, 4, 6) triangles is 3, whose vertices are {4, 6, 12}, {8, 10, 4}, {12, 2, 8}.

So the minimum n is 24.

Second Day

(0800 - 1200; July 29, 2009)

5 The set of the permutation $X = (x_1, x_2, \ldots, x_9)$ of 1, 2, ..., 9 is A. $\forall X \in A$, and let $f(X) = x_1 + 2x_2 + 3x_3 + \cdots + 9x_9$, $M = \{f(X) \mid X \in A\}$. Find the value of $|M|$. (Posed by Xiong Bin)

Solution We prove for $n \geqslant 4$. If the permutations $X_n = (x_1, x_2, \ldots, x_n)$ of $1, 2, \ldots, n$ consist of a set A, and $f(X_n) = x_1 + 2x_2 + 3x_3 + \cdots + nx_n$, $M_n = \{f(X) \mid X \in A\}$, then $|M_n| = \frac{n^3 - n + 6}{6}$.

Using mathematical induction on n, we see that

$$M_n = \left\{\frac{n(n+1)(n+2)}{6}, \frac{n(n+1)(n+2)}{6} + 1, \cdots, \frac{n(n+1)(2n+1)}{6}\right\}.$$

For $n = 4$, by arranging inequality, one can see the smallest number of M is $f(\{4, 3, 2, 1\}) = 20$, and the biggest number is $f(\{1, 2, 3, 4\}) = 30$. Because

$$f(\{3, 4, 2, 1\}) = 21,\ f(\{3, 4, 1, 2\}) = 22,$$
$$f(\{4, 2, 1, 3\}) = 23,\ f(\{3, 2, 4, 1\}) = 24,$$
$$f(\{2, 4, 1, 3\}) = 25,\ f(\{1, 4, 3, 2\}) = 26,$$
$$f(\{1, 4, 2, 3\}) = 27,\ f(\{2, 1, 4, 3\}) = 28,$$
$$f(\{1, 2, 4, 3\}) = 29,$$

we have, $|M_4| = |\{20, 21, \ldots, 30\}| = 11 = \frac{4^3 - 4 + 6}{6}$.

Suppose that the statement is true for $n-1$ $(n \geqslant 5)$. In the case of n, for a permutation $X_{n-1} = (x_1, x_2, \ldots, x_{n-1})$ of $1, 2, \ldots, n-1$, let $x_n = n$; then we get a permutation $(x_1, x_2, \ldots, x_{n-1}, n)$ of $1, 2, \ldots, n$, and so

$$\sum_{k=1}^{n} kx_k = n^2 + \sum_{k=1}^{n-1} kx_k.$$

According to the supposition, the value of $\sum_{k=1}^{n} kx_k$ can be every integer number in the interval

$$\left[n^2 + \frac{(n-1)n(n+1)}{6}, n^2 + \frac{(n-1)n(2n-1)}{6}\right]$$

$$= \left[\frac{n(n^2+5)}{6}, \frac{n(n+1)(2n+1)}{6}\right].$$

Let $x_n = 1$. Then

$$\begin{aligned}\sum_{k=1}^{n} kx_k &= n + \sum_{k=1}^{n-1} kx_k \\ &= n + \sum_{k=1}^{n-1} k(x_k - 1) + \frac{n(n-1)}{2} \\ &= \frac{n(n+1)}{2} + \sum_{k=1}^{n-1} k(x_k - 1).\end{aligned}$$

According to the supposition, the value of $\sum_{k=1}^{n} kx_k$ can be every integer number in the interval

$$\left[\frac{n(n+1)}{2} + \frac{(n-1)n(n+1)}{6}, \frac{n(n+1)}{2} + \frac{n(n-1)(2n-1)}{6}\right]$$

$$= \left[\frac{n(n+1)(n+2)}{6}, \frac{2n(n^2+2)}{6}\right].$$

Because $\frac{2n(n^2+2)}{6} \geqslant \frac{n(n^2+5)}{6}$, according to the supposition the value of $\sum_{k=1}^{n} kx_k$ can be every integer number in the interval

$$\left[\frac{n(n+1)(n+2)}{6}, \frac{n(n+1)(2n+1)}{6}\right].$$

The statement is also true for n. Since

$$\frac{n(n+1)(2n+1)}{6} - \frac{n(n+1)(n+2)}{6} = \frac{n^3-n+6}{6},$$

one can see that $|M_n| = \frac{n^3-n+6}{6}+1.$

In particular, $|M_9| = 121.$

6 Let O, I be the circumcenter and incenter of $\triangle ABC$. Prove that, for an arbitrary point D on the circle O, one can construct a triangle DEF, such that O, I are the circumcenter and incenter of $\triangle DEF$. (Posed by Tao Pingsheng)

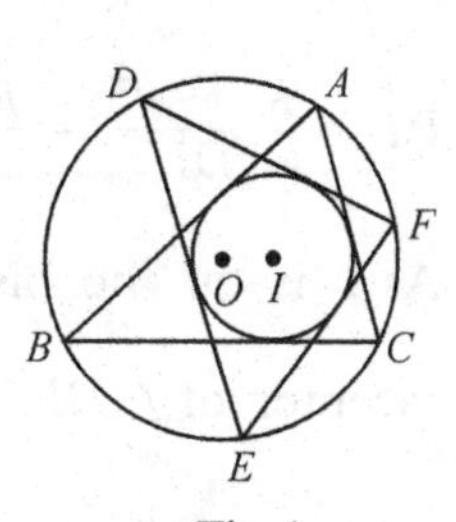

Fig. 1

Solution As shown in Fig. 2, let $OI = d$, R, r be the circumradius and incircle radius of $\triangle ABC$. The point K is the intersection of AI and the circle O; then

$$KI = KB = 2R\sin\frac{\angle BAC}{2},$$

$$AI = \frac{r}{\sin\frac{\angle BAC}{2}}.$$

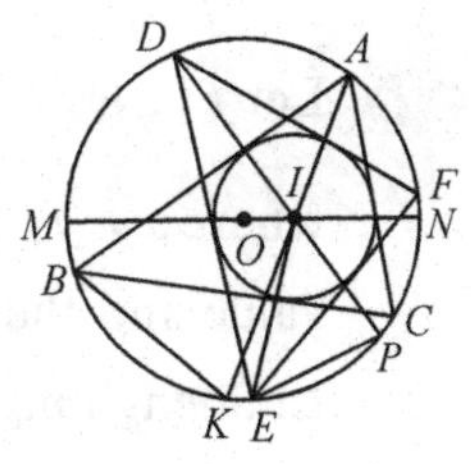

Fig. 2

Let the points M, N be the intersections

of the line OI and the circle O; then

$$(R+d)(R-d) = IM \times IN = AI \times KI = 2Rr,$$

i.e. $R^2 - d^2 = 2Rr$.

Now draw the tangents DE, DF from D to the circle I. The points E, F are on the circle O. Then DI is the bisector of $\angle EDF$. It is enough to prove that EF is tangent to the circle I.

Let P be the intersections of the line DI and the circle O. Then P is the midpoint of the arc EF, and

$$PE = 2R\sin\frac{\angle EDF}{2},\ DI = \frac{r}{\sin\frac{\angle EDF}{2}},$$

$$ID \cdot IP = IM \cdot IN = (R+d)(R-d) = R^2 - d^2,$$

and so

$$PI = \frac{R^2-d^2}{DI} = \frac{R^2-d^2}{r} \cdot \sin\frac{\angle EDF}{2} = 2R\sin\frac{\angle EDF}{2} = PE.$$

As I is on the bisector of $\angle EDF$, we can see that I is the incenter of $\triangle DEF$ $\left[\text{because } \angle PEI = \angle PIE = \frac{1}{2}(180^\circ - \angle EPD) = \frac{1}{2}(180^\circ - \angle DFE) = \frac{\angle EDF + \angle DEF}{2}, \text{ and } \angle PEF = \frac{\angle EDF}{2}; \text{ so } \angle FEI = \frac{\angle DEF}{2}\right]$. EF is tangent to the circle I.

7 Let $f(x, y, z) = \dfrac{x(2y-z)}{1+x+3y} + \dfrac{y(2z-x)}{1+y+3z} + \dfrac{z(2x-y)}{1+z+3x}$, where $x, y, z \geqslant 0$, and $x+y+z=1$. Find the maximum value and the minimum value of $f(x, y, z)$. (Posed by Li Shenghong)

Solution First, prove that $f \leqslant \frac{1}{7}$; when $x = y = z = \frac{1}{3}$, we have $f = \frac{1}{7}$.

Since $f = \sum \frac{x(x+3y-1)}{1+x+3y} = 1 - 2\sum \frac{x}{1+x+3y}$, by Cauchy's inequality

$$\sum \frac{x}{1+x+3y} \geqslant \frac{\left(\sum x\right)^2}{\sum x(1+x+3y)} = \frac{1}{\sum x(1+x+3y)},$$

and

$$\sum x(1+x+3y) = \sum x(2x+4y+z) = 2 + \sum xy \leqslant \frac{7}{3}.$$

So $\sum \frac{x}{1+x+3y} \geqslant \frac{3}{7}$, $f \leqslant 1 - 2 \times \frac{3}{7} = \frac{1}{7}$; $f_{\max} = \frac{1}{7}$; when $x = y = z = \frac{1}{3}$, we have $f = \frac{1}{7}$.

Second, prove that $f \geqslant 0$; when $x = 1$, $y = z = 0$, we have $f = 0$.

In fact, one can see that

$$\begin{aligned} f(x, y, z) &= \frac{x(2y-z)}{1+x+3y} + \frac{y(2z-x)}{1+y+3z} + \frac{z(2x-y)}{1+z+3x} \\ &= xy\left(\frac{2}{1+x+3y} - \frac{1}{1+y+3z}\right) \\ &\quad + xz\left(\frac{2}{1+z+3x} - \frac{1}{1+x+3y}\right) \\ &\quad + yz\left(\frac{2}{1+y+3z} - \frac{1}{1+z+3x}\right) \\ &= \frac{7xyz}{(1+x+3y)(1+y+3z)} \\ &\quad + \frac{7xyz}{(1+z+3x)(1+x+3y)} \\ &\quad + \frac{7xyz}{(1+y+3z)(1+z+3x)} \\ &\geqslant 0. \end{aligned}$$

So $f_{\min} = 0$ and $f_{\max} = \frac{1}{7}$.

8 On a piece of 8×8 graph paper, at least how many grids should be taken off, and we cannot cut out a "T" which has five grids as shown in Fig. 1? (Posed by Sun Wen-Hsien)

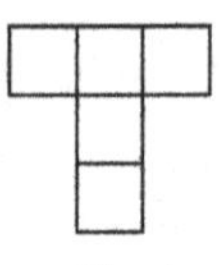
Fig. 1

Solution At least 14 grids should be taken off. An example is given in Fig. 2.

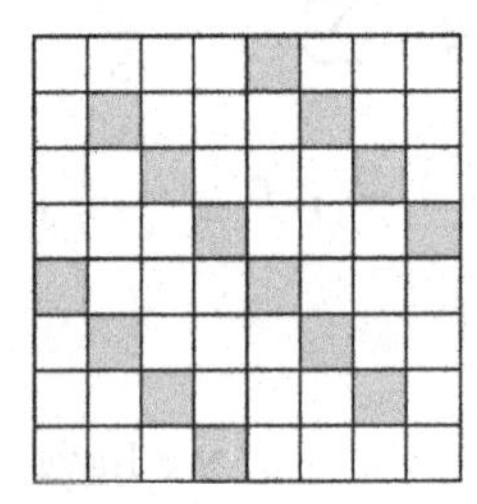
Fig. 2

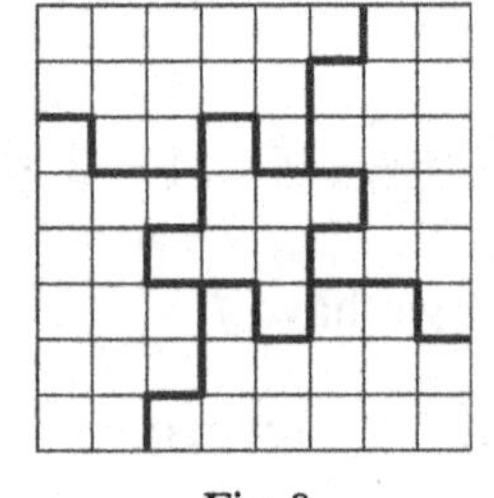
Fig. 3

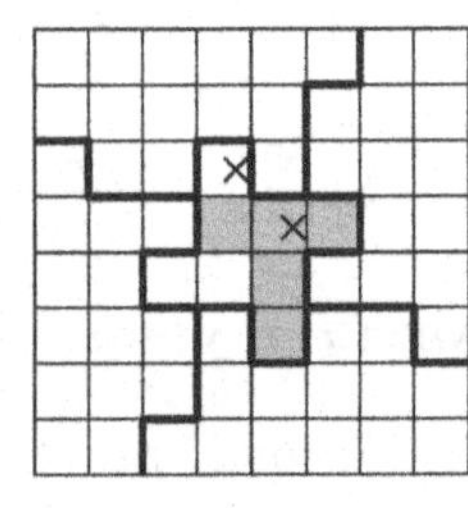
Fig. 4

See Fig. 3: cut the 8×8 graph paper into five areas. We need to take off two grids from the center areas, otherwise we can get a "T" with five grids in the area. Because the two grids with "×" are not equivalent, one is in the corner, and the other is in the center — only one grid is taken off, and we can get a "T" with five grids, as shown in Fig. 4.

For the four corner areas, we prove that at least three grids should be taken off in every area to ensure that we cannot get a "T" with five grids in the area.

For example, in the upper right corner, the "T" below as shown in Fig. 5 should be taken off one grid, and the grids over the "T" should be taken off the "×"

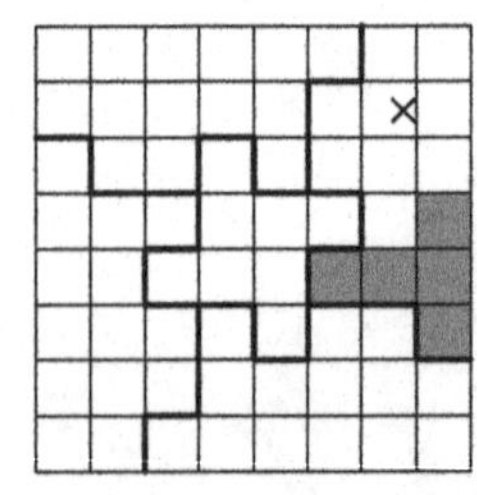
Fig. 5

grid. From the "T" below having five cases, we can check that if only two grids are taken off, we can get a "T" with five grids, which is shown as follows:

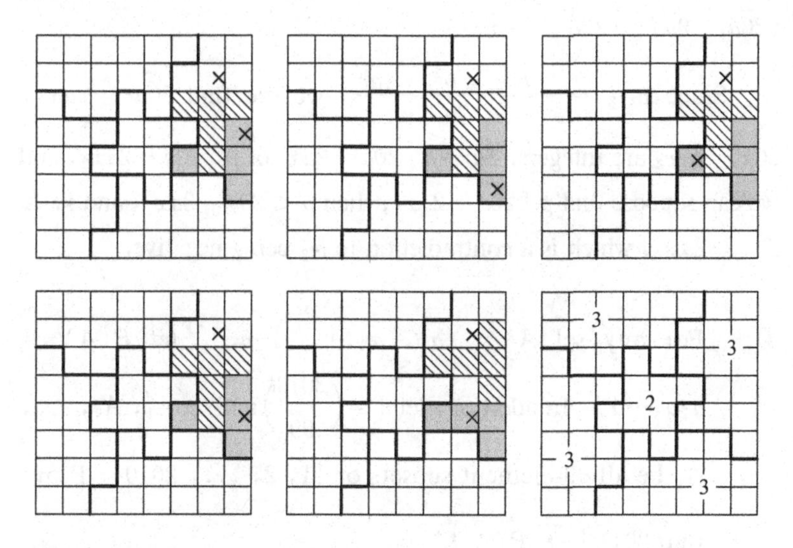

So the answer is $2 + 3 \times 4 = 14$.

2010 (Changhua, Taiwan)

First Day

(0080 - 1200; August 17, 2010)

1 Let $a, b, c \in \{0, 1, 2, \ldots, 9\}$. The quadratic equation $ax^2 + bx + c = 0$ has a rational root. Prove that the three-digit number $\overline{abc}$ is not a prime number.

Solution If $\overline{abc} = p$ is a prime number, and the roots of the equation $f(x) = ax^2 + bx + c = 0$ are rational numbers, then

$b^2 - 4ac$ is a square number, x_1, x_2 are negative, and $f(x) = a(x - x_1)(x - x_2)$.

So $p = f(10) = a(10 - x_1)(10 - x_2)$, and we get $4ap = (20a - 2ax_1)(20a - 2ax_2)$.

Since $x_{1,2} = \dfrac{-b \pm \sqrt{b^2 - 4ac}}{2a}$, we see that $20a - 2ax_1$, $20a - 2ax_2$ are integers. Now $p \mid 20a - 2ax_1$ or $p \mid 20a - 2ax_2$, and we can suppose that $p \mid 20a - 2ax_1$; then $p \leqslant 20a - 2ax_1$, and $4a \geqslant 20a - 2ax_2$, which is a contradiction to x_2 being negative.

2 For any set $A = \{a_1, a_2, \dots, a_m\}$, let $P(A) = a_1 a_2 \cdots a_m$. In addition, let $n = \binom{2010}{99}$ and let $A_1, A_2, \dots, A_n$ be all 99-element subsets of $\{1, 2, \dots, 2010\}$. Prove that $2011 \mid \sum_{i=1}^{n} P(A_i)$.

Solution One can check that 2011 is a prime number. Let

$$f(x) = (x-1)(x-2)\cdots(x-2010) - (x^{2010} + 2010!).$$

For $n \in \{1, 2, \dots, 2010\}$, we have $n^{2010} \equiv 1 \pmod{2011}$ (by Fermat's little theorem), and $2010! \equiv -1 \pmod{2011}$ (by Wilson's theorem), so

$$f(n) \equiv (n-1)(n-2)\cdots(n-2010) \equiv 0 \pmod{2011}.$$

This means that $f(x) \equiv 0 \pmod{2011}$ have 2010 roots, but the degree of $f(x)$ is 2009. Using Langrange's theorem, we can see that the coefficients can all be divided by 2011.

$\sum_{i=1}^{n} P(A_i)$ is the coefficient of x^{1911}, hence $2011 \mid \sum_{i=1}^{n} P(A_i)$.

3 The incircle of the triangle ABC touches BC at D and AB

at F, and intersects the line AD again at H and the line CF again at K. Prove that $\frac{FD \times HK}{FH \times DK} = 3$.

Solution Let $AF = x$, $BF = y$, $CD = z$. Then, by Stewart's theorem,

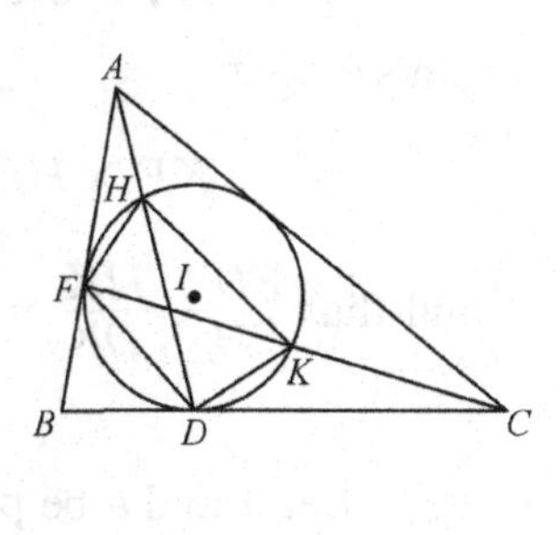

$$AD^2 = \frac{BD}{BC} \times AC^2 + \frac{CD}{BC} \times AB^2 - BD \times DC$$
$$= \frac{y(x+z)^2 + z(x+y)^2}{y+z} - yz$$
$$= x^2 + \frac{4xyz}{y+z}.$$

By the power point theorem,

$$AH = \frac{AF^2}{AD} = \frac{x^2}{AD},$$

$$HD = AD - AH = \frac{AD^2 - x^2}{AD} = \frac{4xyz}{AD(y+z)}.$$

Similarly, we have $KF = \frac{4xyz}{CF(x+y)}$; from $\triangle CDK \backsim \triangle CFD$, there follows $DK = \frac{DF \times CD}{CF} = \frac{DF}{CF} \times z$.

From $\triangle AFH \backsim \triangle ADF$, there follows $FH = \frac{DF \times AF}{AD} = \frac{DF}{AD} \times x$. Using the cosine theorem, we have

$$DF^2 = BD^2 + BF^2 - 2BD \cdot BF \cos B$$
$$= 2y^2\left(1 - \frac{(y+z)^2 + (x+y)^2 - (x+z)^2}{2(x+y)(y+z)}\right)$$
$$= \frac{4xy^2z}{(x+y)(y+z)}.$$

So
$$\frac{KF \times HD}{FH \times DK} = \frac{\frac{4xyz}{CF(x+y)} \times \frac{4xyz}{AD(y+z)}}{\frac{DF}{AD}x \times \frac{DF}{CF}z}$$

$$= \frac{16xy^2z}{DF^2(x+y)(y+z)} = 4.$$

D, K, H, F are concyclic, so by the Ptolemy theorem we can see that

$$KF \times HD = DF \times HK + FH \times DK,$$

and thus $\frac{FD \times HK}{FH \times DK} = 3.$

4 Let a and b be positive integers such that $1 \leqslant a < b \leqslant 100$. If there exists a positive integer k such that $ab \mid (a^k + b^k)$, we say that the pair (a, b) is good. Determine the number of good pairs.

Solution One can see that when $k = 1$, if $a \geqslant 2$, then $ab > a + b$; and if $a = 1$, then $b \nmid (b+1)$. So $k \geqslant 2$.

Now, suppose that $(a, b) = d$, $a = sd$, $b = td$, then $(s, t) = 1$, with $t > 1$. By $std^2 \mid d^k(s^k + t^k)$, it follows that $st \mid d^{k-2}(s^k + t^k)$, as $(st, s^k + t^k) = 1$, we have $st \mid d^{k-2}$, and the prime divisors of st are the divisor of d.

If s or t has a prime divisor p bigger than 11, then $p \mid d$, and so $p^2 \mid a$ or $p^2 \mid b$, but $p^2 > 100$; this is a contradiction. Thus, the prime divisor belongs to $\{2, 3, 5, 7\}$.

Let T be the set of the prime divisors of st. Then $T \neq \{3, 7\}$, otherwise one of a or b is bigger than (or equal to) $7 \times 3 \times 7 > 100$, as similarly, $T \neq \{5, 7\}$. So T is one of the following sets:

$$\{2\}, \{3\}, \{5\}, \{7\}, \{2, 3\}, \{2, 5\}, \{2, 7\}, \{3, 5\}.$$

(1) When $T = \{3, 5\}$, we have $d = 15$, and $s = 3$, $t = 5$, and only one good pair $(a, b) = (45, 75)$;

(2) When $T = \{2, 7\}$, we have $d = 14$, and $(s, t) = (2, 7)$

or (4, 7), and only two good pairs, $(a, b) = (28, 98)$ or $(56, 98)$;

(3) When $T = \{2, 5\}$, we have $d = 10$ or 20, and the coincide $(s, t) = (1, 10), (2, 5), (4, 5), (5, 8)$ or $(2, 5), (4, 5)$. There are six good pairs (a, b);

(4) When $T = \{2, 3\}$, we have $d = 6, 12, 18, 24, 30$, and the coincide $(s, t) = (1, 6), (1, 12), (2, 3), (2, 9), (3, 4), (3, 8), (3, 16), (4, 9), (8, 9), (9, 16)$ or $(1, 6), (2, 3), (3, 4), (3, 8)$ or $(2, 3), (3, 4)$ or $(2, 3), (3, 4)$ or $(2, 3)$. There are 19 good pairs (a, b);

(5) When $T = \{7\}$, we have $(s, t) = (1, 7)$, $d = 7$ or 14. There are two good pairs (a, b);

(6) When $T = \{5\}$, we have $(s, t) = (1, 5)$, $d = 5, 10, 15$ or 20. There are four good pairs (a, b);

(7) When $T = \{3\}$, we have $(s, t) = (1, 3), (1, 9)$ or $(1, 27)$, and the coincide $d \in \{3, 6, \ldots, 33\}, \{3, 6, 9\}$ or $\{3\}$. There are 15 good pairs (a, b);

(8) When $T = \{2\}$, we have $(s, t) = (1, 2), (1, 4), (1, 8), (1, 16)$ or $(1, 32)$, and the coincide $d \in \{2, 4, \ldots, 50\}, \{2, 4, \ldots, 24\}, \{2, 4, \ldots, 12\}, \{2, 4, 6\}$ or $\{2\}$. There are 47 good pairs (a, b).

So we have $1 + 2 + 6 + 19 + 2 + 4 + 15 + 47 = 96$ good pairs (a, b) satisfied.

Second Day

(0080 – 1200; August 18, 2010)

5 ABC is a triangle with a right angle at C. M_1 and M_2 are two arbitrary points inside ABC, and M is the midpoint of M_1M_2. The extensions of BM_1, BM and BM_2 intersect

AC at N_1, N and N_2 respectively.

Prove that $\frac{M_1N_1}{BM_1}+\frac{M_2N_2}{BM_2}\geqslant 2\frac{MN}{BM}$.

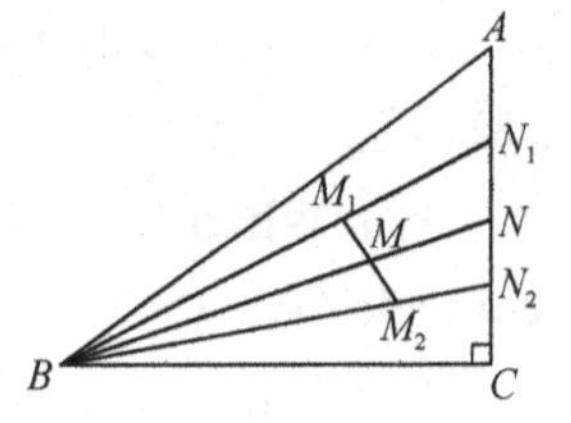

Solution Let H_1, H_2, H be the projection from M_1, M_2, M to the line BC. Then

$$\frac{M_1N_1}{BM_1}=\frac{H_1C}{BH_1},$$

$$\frac{M_2N_2}{BM_2}=\frac{H_2C}{BH_2},$$

$$\frac{MN}{BM}=\frac{HC}{BH}=\frac{H_1C+H_2C}{BH_1+BH_2}.$$

Now suppose that $BC=1$, $BH_1=x$, $BH_2=y$. Then

$$\frac{M_1N_1}{BM_1}=\frac{H_1C}{BH_1}=\frac{1-x}{x},$$

$$\frac{M_2N_2}{BM_2}=\frac{H_2C}{BH_2}=\frac{1-y}{y},$$

$$\frac{MN}{BM}=\frac{HC}{BH}=\frac{1-x+1-y}{x+y}.$$

So it is enough to prove that

$$\frac{1-x}{x}+\frac{1-y}{y}\geqslant 2\frac{1-x+1-y}{x+y},$$

which is equivalent to $\frac{1}{x}+\frac{1}{y}\geqslant\frac{4}{x+y}$, i.e. $(x-y)^2\geqslant 0$.

This is obviously true.

6 Let $\mathbf{N}^*$ be the set of positive integers. Define $a_1=2$, and for $n=1, 2, \ldots$,

$$a_{n+1}=\min\left\{\lambda\,\middle|\,\frac{1}{a_1}+\frac{1}{a_2}+\cdots+\frac{1}{a_n}+\frac{1}{\lambda}<1,\ \lambda\in\mathbf{N}^*\right\}.$$

Prove that $a_{n+1} = a_n^2 - a_n + 1$ for $n = 1, 2, \ldots$.

Solution Since $a_1 = 2$, $a_2 = \min\left\{\lambda \,\middle|\, \frac{1}{a_1} + \frac{1}{\lambda} < 1, \lambda \in \mathbf{N}^*\right\}$, from $\frac{1}{a_1} + \frac{1}{\lambda} < 1$ we have $\frac{1}{\lambda} < 1 - \frac{1}{2} = \frac{1}{2}$, $\lambda > 2$ and so $a_2 = 3$.

This means that when $n = 1$, the conclusion is right.

Suppose that for $n \leqslant k - 1$ $(k \geqslant 2)$, the conclusions are right, for $n = k$, from

$$a_{k+1} = \min\left\{\lambda \,\middle|\, \frac{1}{a_1} + \frac{1}{a_2} + \cdots + \frac{1}{a_k} + \frac{1}{\lambda} < 1, \lambda \in \mathbf{N}^*\right\}.$$

As $\frac{1}{a_1} + \frac{1}{a_2} + \cdots + \frac{1}{a_k} + \frac{1}{\lambda} < 1$, i.e.

$$0 < \frac{1}{\lambda} < 1 - \left(\frac{1}{a_1} + \frac{1}{a_2} + \cdots + \frac{1}{a_k}\right),$$

$$\lambda > \frac{1}{1 - \frac{1}{a_1} - \frac{1}{a_2} - \cdots - \frac{1}{a_k}}.$$

Now we prove that $\frac{1}{1 - \frac{1}{a_1} - \frac{1}{a_2} - \cdots - \frac{1}{a_k}} = a_k(a_k - 1)$.

By the supposition, for $2 \leqslant n \leqslant k$, $a_n = a_{n-1}(a_{n-1} - 1) + 1$; then

$$\frac{1}{a_n - 1} = \frac{1}{a_{n-1}(a_{n-1} - 1)} = \frac{1}{a_{n-1} - 1} - \frac{1}{a_{n-1}}.$$

So $\frac{1}{a_{n-1}} = \frac{1}{a_{n-1} - 1} - \frac{1}{a_n - 1}$, and $\sum_{i=2}^{k} \frac{1}{a_{i-1}} = 1 - \frac{1}{a_k - 1}$, i.e.

$$\sum_{i=1}^{k} \frac{1}{a_i} = 1 - \frac{1}{a_k - 1} + \frac{1}{a_k} = 1 - \frac{1}{a_k(a_k - 1)},$$

which means that $\frac{1}{1 - \frac{1}{a_1} - \frac{1}{a_2} - \cdots - \frac{1}{a_k}} = a_k(a_k - 1)$; then

$$a_{k+1} = \min\left\{\lambda \,\middle|\, \frac{1}{a_1} + \frac{1}{a_2} + \cdots + \frac{1}{a_k} + \frac{1}{\lambda} < 1,\ \lambda \in \mathbf{N}^*\right\}$$
$$= a_k(a_k - 1) + 1.$$

We get the conclusion for all n: $a_{n+1} = a_n^2 - a_n + 1$.

7 Let n be a positive integer. The real numbers a_1, a_2, ..., a_n and r_1, r_2, ..., r_n are such that $a_1 \leqslant a_2 \leqslant \cdots \leqslant a_n$ and $0 \leqslant r_1 \leqslant r_2 \leqslant \cdots \leqslant r_n$.

Prove that $\sum_{i=1}^{n}\sum_{j=1}^{n} a_i a_j \min(r_i, r_j) \geqslant 0$.

Solution Let

$$A_1 = \begin{pmatrix} a_1a_1r_1 & a_1a_2r_1 & a_1a_3r_1 & \cdots & a_1a_nr_1 \\ a_2a_1r_1 & a_2a_2r_2 & a_2a_3r_2 & \cdots & a_2a_nr_2 \\ a_3a_1r_1 & a_3a_2r_2 & a_3a_3r_3 & \cdots & a_3a_nr_3 \\ \vdots & \vdots & \vdots & & \vdots \\ a_na_1r_1 & a_na_2r_2 & a_na_3r_3 & \cdots & a_na_nr_n \end{pmatrix}$$

Since

$$\sum_{i=1}^{n}\sum_{j=1}^{n} a_i a_j \min(r_i, r_j)$$
$$= \sum_{j=1}^{n} a_1 a_j \min(r_1, r_j) + \sum_{j=1}^{n} a_2 a_j \min(r_2, r_j) + \cdots$$
$$+ \sum_{j=1}^{n} a_k a_j \min(r_k, r_j) + \cdots + \sum_{j=1}^{n} a_n a_j \min(r_n, r_j);$$

whose k term is

$$\sum_{j=1}^{n} a_k a_j \min(r_k, r_j)$$
$$= a_ka_1r_1 + a_ka_2r_2 + \cdots + a_ka_kr_k + a_ka_{k+1}r_k + \cdots + a_ka_nr_k,$$

i.e. the sum of the elements from the k row of the graph, $k =$

1, 2, ..., n, we find that $\sum_{i=1}^{n}\sum_{j=1}^{n} a_i a_j \min(r_i, r_j)$ is the sum of all elements in A_1.

On the other hand, the sum can be calculated as follows. First, count the sum of elements from the first row and the first column of A_1, and let the other elements constitute a graph A_2. Then count the sum of elements from the first row and the first column of A_2, ... So

$$\begin{aligned}
&\sum_{i=1}^{n}\sum_{j=1}^{n} a_i a_j \min(r_i, r_j) \\
&= \sum_{k=1}^{n} r_k (a_k^2 + 2a_k(a_{k+1} + a_{k+2} + \cdots + a_n)) \\
&= \sum_{k=1}^{n} r_k \left(\left(a_k + \sum_{i=k+1}^{n} a_i \right)^2 - \left(\sum_{i=k+1}^{n} a_i \right)^2 \right) \\
&= \sum_{k=1}^{n} r_k \left(\left(\sum_{i=k}^{n} a_i \right)^2 - \left(\sum_{i=k+1}^{n} a_i \right)^2 \right) \\
&= r_1 \left(\sum_{i=1}^{n} a_i \right)^2 + r_2 \left(\sum_{i=2}^{n} a_i \right)^2 + r_3 \left(\sum_{i=3}^{n} a_i \right)^2 + \cdots \\
&\quad + r_n \left(\sum_{i=n}^{n} a_i \right)^2 \text{ (where } r_0 = 0 \text{)} \\
&\quad - r_1 \left(\sum_{i=2}^{n} a_i \right)^2 - r_2 \left(\sum_{i=3}^{n} a_i \right)^2 - \cdots - r_{n-1} \left(\sum_{i=n}^{n} a_i \right)^2 \\
&= \sum_{k=1}^{n} (r_k - r_{k-1}) \left(\sum_{i=k}^{n} a_i \right)^2 \geqslant 0.
\end{aligned}$$

8 $A_1, A_2, \ldots, A_8$ are fixed points on a circle. Determine the smallest positive integer n such that among any n triangles with these eight points as vertices, two of them will have a common side.

Solution First, we prove that if there are r triangles such that every two of them have no common side, then the maximum of r is 8.

There are $\binom{8}{2} = 28$ chords whose endpoints are from the 8 points.

If every chord belongs to only one triangle, then the maximum $r \leqslant \left[\frac{28}{3}\right] = 9$. But if there 9 triangles such that every 2 of them have no common side, then there are 27 vertices of this triangle, and there is a point belonging to 4 triangles (suppose that it is A_8), whose opposite sides are chords with endpoints of $A_1, A_2, \ldots, A_7$, and so there is a point A_k that is the endpoint of two triangles, and the two triangles have a common side $A_8 A_k$, so $r \leqslant 8$.

On the other hand, the following example shows that r can be 8: from the picture, the triangle (1, 2, 8), (1, 3, 6), (1, 4, 7), (2, 3, 4), (2, 5, 7), (3, 5, 8), (4, 5, 6), (6, 7, 8) is satisfied. So the maximum of r is 8.

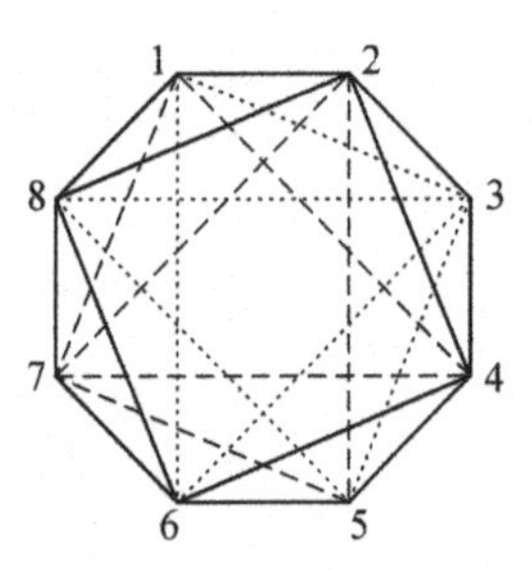

Back to the problem, the minimum n is $8+1=9$.

International Mathematical Olympiad

2009 (Bremen, Germany)

First Day

(0900 - 1330; July 15, 2010)

1 Let n be a positive integer and let $a_1, a_2, \ldots, a_k (k \geqslant 2)$ be distinct integers in the set $\{1, \ldots, n\}$ such that n divides $a_i(a_{i+1} - 1)$ for $i = 1, \ldots, k-1$. Prove that n does not divide $a_k(a_1 - 1)$. (Posed by Australia)

Proof We first prove by induction that for any integer $2 \leqslant i \leqslant k$, $n \mid a_1(a_i - 1)$.

If $i=2$, we have $n \mid a_1(a_2-1)$ by assumption. Suppose that $n \mid a_1(a_i-1)$ for some $2 \leqslant i \leqslant k-1$; then

$$n \mid a_1(a_i-1)(a_{i+1}-1).$$

By assumption, we have $n \mid a_i(a_{i+1}-1)$, so $n \mid a_1a_i(a_{i+1}-1)$, and therefore

$$n \mid (a_1a_i(a_{i+1}-1)-a_1(a_i-1)(a_{i+1}-1)),$$

i.e. $n \mid a_1(a_{i+1}-1)$. By the method of induction, $n \mid a_1(a_i-1)$ holds for every integer $2 \leqslant i \leqslant k$. In particular, we have $n \mid a_1(a_k-1)$.

Since $a_1(a_k-1)-a_k(a_1-1)=a_k-a_1$ is not divisible by n (because a_1 and a_k are distinct elements in $\{1, \ldots, n\}$), we conclude that $a_k(a_1-1)$ is not divisible by n.

2 Let ABC be a triangle with circumcenter O. The points P and Q are interior points of the sides CA and AB, respectively. Let K, L and M be the midpoints of the segments BP, CQ and PQ, respectively, and let Γ be the circle passing through K, L and M. Suppose that the line PQ is tangent to the circle Γ. Prove that $OP=OQ$. (Posed by Russia)

Proof Clearly, the line PQ touches the circle Γ at the point M. By the theorem of the tangent chord angle, we have $\angle QMK=\angle MLK$. Since the points K, M are the midpoints of the segments BP and PQ respectively, $KM \parallel BQ$, we get $\angle QMK=\angle AQP$. Thus, $\angle MLK=\angle AQP$, and similarly $\angle MKL=\angle APQ$. As a result, we see that $\triangle MKL$ and $\triangle APQ$ are similar, and thus

$$\frac{MK}{ML}=\frac{AP}{AQ}.$$

Since the points K, L and M are the midpoints of the segments BP, CQ and PQ respectively, we have

$$KM = \frac{1}{2}BQ, \ LM = \frac{1}{2}CP.$$

Plugging this into the previous equation, we get $\frac{BQ}{CP} = \frac{AP}{AQ}$, i.e.

$$AP \times CP = AQ \times BQ.$$

By the power law of a point, we have

$$OP^2 = OA^2 - AP \times CP = OA^2 - AQ \times BQ = OQ^2.$$

Therefore $OP = OQ$, as required.

3 Suppose that s_1, s_2, s_3, ... is a strictly increasing sequence of positive integers such that the subsequences s_{s_1}, s_{s_2}, s_{s_3}, ... and s_{s_1+1}, s_{s_2+1}, s_{s_3+1}, ... are both arithmetic progressions. Prove that the sequence s_1, s_2, s_3, ... is itself an arithmetic progression. (Posed by America)

Proof It follows easily from assumption that s_{s_1}, s_{s_2}, s_{s_3}, ... and s_{s_1+1}, s_{s_2+1}, s_{s_3+1}, ... are both strictly increasing sequences of positive integers.

Assume that $s_{s_k} = a + (k-1)d_1$, $s_{s_k+1} = b + (k-1)d_2$, $k = 1, 2, \ldots$, where a, b, d_1, d_2 are positive integers. Since $s_k < s_k + 1 \leqslant s_{k+1}$, by the monotonicity of the sequence $\{s_n\}$ we have

$$s_{s_k} < s_{s_k+1} \leqslant s_{s_{k+1}},$$

i.e.

$$a+(k-1)d_1<b+(k-1)d_2\leqslant a+kd_1,$$

or, equivalently,

$$a-b<(k-1)(d_2-d_1)\leqslant a+d_1-b.$$

As k is arbitrary, we must have $d_2-d_1=0$, i.e. $d_2=d_1$, denoted by d for both d_1 and d_2. Let $b-a=c\in\mathbf{N}^*$. If $d=1$, by the monotonicity of $\{s_n\}$ we have $s_{s_{k+1}}=s_{s_k}+1\leqslant s_{s_k+1}$, and thus $s_{k+1}\leqslant s_k+1$.

Since $s_{k+1}>s_k$, we get $s_{k+1}=s_k+1$, i. e. $\{s_n\}$ is an arithmetic progression, which is what we want. In what follows, we assume that $d>1$.

We shall prove that $s_{k+1}-s_k=c$ for any positive integer k. Supposing on the contrary that it is not true, we discuss two cases.

Case 1: There exists a positive integer k such that $s_{k+1}-s_k<c$. Since $s_{k+1}-s_k$ are positive integers, we may assume that $s_{i+1}-s_i=c_0$ attains the minimum value for some i; then

$$\begin{aligned}s_{a+id}-s_{a+(i-1)d+1}&=s_{s_{s_i+1}}-s_{s_{s_i}+1}\\&=(a+(s_{i+1}-1)d)-(b+(s_i-1)d)\\&=c_0d-c.\end{aligned}\qquad①$$

On the other hand, since

$$(a+id)-(a+(i-1)d+1)=d-1,$$

we have

$$s_{a+id}-s_{a+(i-1)d+1}\geqslant c_0(d-1)$$

(here we have used the minimality of c_0). Marking a comparison with ①, we have $c_0\geqslant c$, a contradiction.

Case 2: There exists a positive integer k such that $s_{k+1}-s_k>c$. Since $s_{k+1}-s_k$ are integers, and for any k,

$$s_{k+1} - s_k \leqslant s_{s_{k+1}} - s_{s_k} = d,$$

we may assume that $s_{j+1} - s_j = c_1$ attains the maximum value for some j; then

$$\begin{aligned} s_{a+jd} - s_{a+(j-1)d+1} &= s_{s_{s_{j+1}}} - s_{s_{s_j}+1} \\ &= (a + (s_{j+1} - 1)d) - (b + (s_j - 1)d) \\ &= c_1 d - c. \end{aligned} \qquad ②$$

On the other hand, since

$$(a + jd) - (a + (j - 1)d + 1) = d - 1,$$

we have

$$s_{a+jd} - s_{a+(j-1)d+1} \leqslant c_1(d - 1)$$

(here we have used the maximality of c_1). Marking a Comparison with ②, we have $c_1 \leqslant c$, a contradiction.

We have verified that $s_{k+1} - s_k = c$ for any positive integer k, i.e. $\{s_n\}$ is an arithmetic progression.

Second Day

(0900 - 1330; July 16, 2010)

4 Let ABC be a triangle with $AB = AC$. The angle bisectors of $\angle CAB$ and $\angle ABC$ meet the sides BC and CA at D and E respectively. Let K be the incenter of the triangle ADC. Suppose that $\angle BEK = 45°$. Find all possible values of $\angle CAB$. (Posed by Belgium)

Solution Since AD and BE are the angle bisectors of $\angle CAB$ and $\angle ABC$, their intersection point is the incenter I. Join CI; then CI is the angle bisector of $\angle ACB$. Since K is the incenter of $\triangle ADC$, K lies on the segment CI.

Let $\angle BAC = \alpha$, since $AB = AC$, $AD \perp BC$, we have

$$\angle ABC = \angle ACB = 90^\circ - \frac{\alpha}{2}.$$

As BI and CI bisect $\angle ABC$ and $\angle ACB$ respectively, we get

$$\angle ABI = \angle IBC = \angle ACI = \angle ICB = 45^\circ - \frac{\alpha}{4}.$$

Thus,

$$\angle EIC = \angle IBC + \angle ICB = 90^\circ - \frac{\alpha}{2},$$

$$\angle IEC = \angle BAE + \angle ABE = 45^\circ + \frac{3\alpha}{4}.$$

So we have

$$\frac{IK}{KC} = \frac{S_{\triangle IEK}}{S_{\triangle EKC}} = \frac{\frac{1}{2}IE \times EK \times \sin\angle IEK}{\frac{1}{2}EC \times EK \times \sin\angle KEC} = \frac{\sin 45^\circ}{\sin\frac{3\alpha}{4}} \times \frac{IE}{EC} = \frac{\sin 45^\circ}{\sin\frac{3\alpha}{4}} \times \frac{\sin\left(45^\circ - \frac{\alpha}{4}\right)}{\sin\left(90^\circ - \frac{\alpha}{2}\right)}.$$

On the other hand, since K is the incenter of $\triangle ADC$, DK bisects $\angle IDK$. It follows from the property of the angle bisector that

$$\frac{IK}{KC} = \frac{ID}{DC} = \tan\angle ICD = \frac{\sin\left(45^\circ - \frac{\alpha}{4}\right)}{\cos\left(45^\circ - \frac{\alpha}{4}\right)}.$$

Thus,

$$\frac{\sin 45^\circ}{\sin \frac{3\alpha}{4}} \times \frac{\sin\left(45^\circ - \frac{\alpha}{4}\right)}{\sin\left(90^\circ - \frac{\alpha}{2}\right)} = \frac{\sin\left(45^\circ - \frac{\alpha}{4}\right)}{\cos\left(45^\circ - \frac{\alpha}{4}\right)}.$$

Removing the denominators, we have

$$2\sin 45^\circ \cos\left(45^\circ - \frac{\alpha}{4}\right) = 2\sin \frac{3\alpha}{4} \cos \frac{\alpha}{2}.$$

By the product-to-sum identities, we get

$$\sin\left(90^\circ - \frac{\alpha}{4}\right) + \sin \frac{\alpha}{4} = \sin \frac{5\alpha}{4} + \sin \frac{\alpha}{4},$$

or, equivalently,

$$\sin\left(90^\circ - \frac{\alpha}{4}\right) = \sin \frac{5\alpha}{4}.$$

Since $0 < \alpha < 180^\circ$, $\sin\left(90^\circ - \frac{\alpha}{4}\right) > 0$, we have $\sin \frac{5\alpha}{4} > 0$, i.e. $0 < \frac{5\alpha}{4} < 180^\circ$. It follows that either

$$90^\circ - \frac{\alpha}{4} = \frac{5\alpha}{4} \Rightarrow \alpha = 60^\circ$$

or

$$90^\circ - \frac{\alpha}{4} = 180^\circ - \frac{5\alpha}{4} \Rightarrow \alpha = 90^\circ.$$

When $\alpha = 60^\circ$, it is easy to see that $\triangle IEC \cong \triangle IDK$, so

$$\triangle IEK \cong \triangle IDK,$$

and therefore

$$\angle BEK = \angle IDK = 45^\circ.$$

When $\alpha = 90^\circ$,

$$\angle EIC = 90^\circ - \frac{\alpha}{2} = 45^\circ = \angle KDC.$$

Since $\angle IEC = \angle DCK$, $\triangle ICE$ and $\triangle DCK$ are similar, which implies that

$$IC \times KC = DC \times EC.$$

It follows that $\triangle IDC$ and $\triangle EKC$ are similar, so $\angle EKC = \angle IDC = 90^\circ$, and hence

$$\angle BEK = 180^\circ - \angle EIK - \angle EKI = 45^\circ.$$

Combining the above arguments, we conclude that all possible values of $\angle CAB$ are 60° and 90°.

5 Determine all functions f from the set of positive integers to the set of positive integers such that, for all positive integers a and b, there exists a nondegenerate triangle with sides of lengths a, $f(b)$ and $f(b+f(a)-1)$. (A triangle is nondegenerate if its vertices are not collinear.) (Posed by France)

Solution The only f with the required property is $f(n) = n$ for all $n \in \mathbf{N}^*$.

By assumption and discreteness of integers, we see that for any positive integers a, b,

$$f(b) + f(b+f(a)-1) - 1 \geqslant a, \quad ①$$

$$f(b) + a - 1 \geqslant f(b+f(a)-1), \quad ②$$

$$f(b+f(a)-1) + a - 1 \geqslant f(b). \quad ③$$

Setting $a = 1$ in ② and ③, we have $f(b) = f(b+f(1)-1)$ for any $b \in \mathbf{N}^*$.

If $f(1) \neq 1$, then the above equality implies that f is periodic. Since f is defined on positive integers, f is bounded. Let positive integer M be such that $M \geqslant f(n)$ for any positive integer n (i.e. M is an upper bound for f). Putting $a = 2M$ in ① results in a contradiction. Thus, $f(1) = 1$.

Setting $b = 1$ in ① and ②, we have $f(f(n)) = n$ for all $n \in \mathbf{N}^*$.

If there exist some $t \in \mathbf{N}^*$ such that $f(t) < t$, then $t \geqslant 2$ and $f(t) \leqslant t - 1$. Setting $a = f(t)$ in ②, we have

$$\begin{aligned} f(b+t-1) &= f(b+f(a)-1) \\ &\leqslant f(b)+a-1 \leqslant f(b)+t-2. \end{aligned}$$

Let $M = (t-1) \times \max\limits_{1 \leqslant i \leqslant t-1} f(i)$. For any integer $n > M$, denote by n_0 the unique positive integer satisfying

$$1 \leqslant n_0 \leqslant t-1,\ n_0 \equiv n \pmod{t-1}.$$

Then

$$f(n) \leqslant f(n_0) + \frac{t-2}{t-1}(n-n_0) \leqslant \frac{M}{t-1} + \frac{(t-2)n}{t-1} < n.$$

Therefore, $f(n) < n$ whenever integer $n > M$.

Now choose $n_1 \in \mathbf{N}^*$ such that $n_1 > M$ and n_1 is not equal to any of $f(1), f(2), \ldots, f(M)$. Since $f(f(n_1)) = n_1$, it follows that $f(n_1) > M$, and hence

$$n_1 > f(n_1) > f(f(n_1)) = n_1,$$

a contradiction. As a result, we have $f(t) \geqslant t$ for any $t \in \mathbf{N}^*$. Then $t = f(f(t)) \geqslant f(t) \geqslant t$, and all inequalities become equalities, i.e. $f(n) = n$ for any $n \in \mathbf{N}^*$.

It is not hard to check that $f(n) = n$ for all $n \in \mathbf{N}^*$ satisfies the required property, thus completing our conclusion that the only solution to the problem is $f(n) = n$ for all $n \in \mathbf{N}^*$.

6 Let $a_1, a_2, \ldots, a_n$ be distinct positive integers and let M be a set of $n-1$ positive integers not containing $s = a_1 + a_2 + \cdots + a_n$. A grasshopper is to jump along the real axis, starting at the point 0 and making n jumps to the right with lengths $a_1, a_2, \ldots, a_n$ in some order. Prove that the order can be chosen in such a way that the grasshopper never lands on any point in M. (Posed by Russia)

Proof We proceed by induction on n. When $n = 1$, M is empty, and the result is obvious.

When $n = 2$, M does not contain at least one of a_1, a_2, and the grasshopper can first jump the number that is not in M, then the other number.

Let $m \geqslant 3$ and assume that the result is true for any $n < m$, $n \in \mathbf{N}^*$. We shall prove that the result also holds for $n = m$. Suppose on the contrary that there exist m distinct positive integers $a_1, a_2, \ldots, a_m$ and a set M of $m-1$ numbers, such that the grasshopper cannot finish the jumping as required in the problem.

Suppose that the grasshopper can jump at most k steps to the right with lengths of distinct numbers in $a_1, a_2, \ldots, a_m$, such that the landing points are never in M. Since M contains only $m-1$ numbers, the grasshopper can always take its first step. If the grasshopper can jump $m-1$ steps, it can also jump the last step since M does not contain $a_1 + a_2 + \cdots + a_n$, and therefore $1 \leqslant k \leqslant m-2$.

We choose such k steps with a minimum total length (if there is a tie, just choose any one that attains the minimum), denote the lengths of these k steps by $b_1, b_2, \ldots, b_k$, and let $b_{k+1}, b_{k+2}, \ldots, b_m$ be the remaining numbers in $a_1, a_2, \ldots, a_m$ in increasing order. Clearly,

$$\{a_1, a_2, \ldots, a_m\} = \{b_1, b_2, \ldots, b_m\}.$$

By assumption, for any $k+1 \leqslant j \leqslant m$, $b_1 + b_2 + \cdots + b_k + b_j$ belongs to M. Thus, there are at least $m-k$ elements of M that are greater than or equal to $b_1 + b_2 + \cdots + b_{k+1}$, and hence there are at most $k-1$ elements of M that are less than $b_1 + b_2 + \cdots + b_{k+1}$.

Let

$$A = \{b_i \mid 1 \leqslant i \leqslant k+1,\ i \in \mathbf{Z},\ b_1 + b_2 + \cdots + b_{k+1} - b_i \notin M\}.$$

Then $b_{k+1} \in A$. For any $b_i \in A$, if we remove b_i from $b_1, b_2, \ldots, b_{k+1}$, the sum of the remaining k numbers is not in M, and since

$$|M \cap \{1, 2, \ldots, b_1 + b_2 + \cdots + b_{k+1} - b_i\}| \leqslant k-1,$$

and $k < m$, by the inductive hypothesis there exists an ordering $c_1, c_2, \ldots, c_k$ of these remaining numbers, such that

$$c_1, c_1 + c_2, \ldots, c_1 + c_2 + \cdots + c_k$$

are not in M. Thus, $c_1, c_2, \ldots, c_k$ is also a possible length of k jumps, and by the choice of $b_1, b_2, \ldots, b_k$ we have

$$b_1 + b_2 + \cdots + b_k \leqslant c_1 + c_2 + \cdots + c_k,$$

i.e. $b_i \leqslant b_{k+1}$ for any $b_i \in A$.

Let $A = \{x_1, x_2, \ldots, x_t\}$, where $x_1 < x_2 < \cdots < x_t = b_{k+1}$. By definition of A, we see that there are at least $k+1-t$ numbers of M that are less than $b_1 + b_2 + \cdots + b_{k+1}$. Since for any $k+1 \leqslant j \leqslant m$, $b_1 + b_2 + \cdots + b_k + b_j$ belongs to M, there are at least $m-k$ numbers of M in the interval

$$[b_1 + b_2 + \cdots + b_{k+1},\ b_1 + b_2 + \cdots + b_k + b_m].$$

On the other hand, for any $x_i \in A$, the number

$$b_1 + b_2 + \cdots + b_{k+1} - x_i + b_m$$

belongs to M (otherwise, since $k \leqslant m-2$, b_m is different from $b_1, b_2, \ldots, b_{k+1}$, and the grasshopper can take $k+1$ jumps). For $1 \leqslant i \leqslant t-1$, the numbers

$$b_1 + b_2 + \cdots + b_{k+1} - x_i + b_m$$

are pairwise distinct and greater than $b_1 + b_2 + \cdots + b_k + b_m$, and hence there are at least $t-1$ numbers of M greater than $b_1 + b_2 + \cdots + b_k + b_m$. Summing up, M contains at least

$$(k+1-t) + (m-k) + (t-1) = m$$

numbers, which contradicts the fact that M contains only $m-1$ numbers.

Thus, our assumption is false, and the result also holds for $n = m$. By the method of mathematical induction, the result holds for every positive integer n.

2010 (Astana, Kazakhstan)

First Day

(0900-1330; July 7, 2010)

1 Determine all functions $f: \mathbf{R} \to \mathbf{R}$ such that the equality

$$f([x]y) = f(x)[f(y)] \qquad ①$$

holds for all $x, y \in \mathbf{R}$. (Here $[z]$ denotes the greatest integer less than or equal to z.)(Posed by France)

Solution The answer is $f(x) = C$ (a constant function), where $C = 0$ or $1 \leqslant C < 2$.

Setting $x = 0$ in ①, we have

$$f(0) = f(0)[f(y)] \qquad ②$$

for all $y \in \mathbf{R}$. We discuss the following two cases:

(1) When $f(0) \neq 0$, it follows from ② that $[f(y)] = 1$ for all $y \in \mathbf{R}$. Then ① becomes $f([x]y) = f(x)$, and setting $y = 0$ we get $f(x) = f(0) = C \neq 0$.

Since $[f(y)] = 1 = [C]$, we have $1 \leqslant C < 2$.

(2) When $f(0) = 0$, assume that there exists $0 < \alpha < 1$ such that $f(\alpha) \neq 0$. We set $x = \alpha$ in ①, and then $f(y) = 0$ for all $y \in \mathbf{R}$, which contradicts $f(\alpha) \neq 0$.

We now assume that $f(\alpha) = 0$ for all $0 \leqslant \alpha < 1$. For any real number z, there is an integer N with $\alpha = \frac{z}{N} \in [0, 1)$, and it follows from ① that

$$f(z) = f([N]\alpha) = f(N)[f(\alpha)] = 0$$

for all $z \in \mathbf{R}$.

It is easy to check that $f(x) = C$ (a constant function) with $C = 0$ or $1 \leqslant C < 2$ satisfies the required property of the problem.

2 Let I be the incenter of the triangle ABC and let Γ be its circumcircle. Let the line AI intersect Γ again at D. Let E be a point on the arc $\widehat{BDC}$ and F a point on the side BC such that

$$\angle BAF = \angle CAE < \frac{1}{2}\angle BAC.$$

Finally, let G be the midpoint of the segment IF. Prove that the lines DG and EI intersect on Γ. (Posed by

Hongkong)

Proof Let I_a be the center of the described circle of the triangle ABC with respect to the side BC. Then D is the midpoint of the segment II_a, $DG \parallel I_a F$ and $\angle GDA = \angle FI_aA$.

To prove that the intersection point P of the lines DG and EI lies on Γ, i.e. A, E, D, P are concyclic, is equivalent to proving that $\angle GDA = \angle IEA$, or similarly $\angle FI_aA = \angle IEA$.

By the assumption $\angle BAF = \angle CAE$, we have $\triangle ABF \backsim \triangle AEC$, and hence

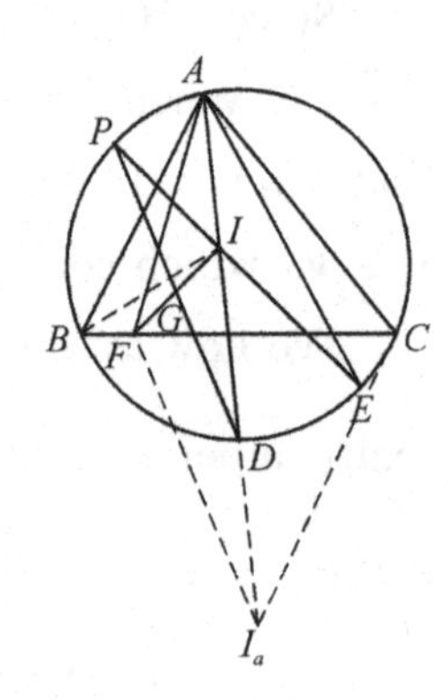

$$\frac{AF}{AB} = \frac{AC}{AE}. \quad ①$$

Since

$$\angle AIB = \angle C + \frac{1}{2}(\angle A + \angle B),$$

$$\angle ACI_a = \angle C + \frac{1}{2}(\angle A + \angle B),$$

we have $\triangle ABI \backsim \triangle AI_aC$, and thus

$$\frac{AI}{AB} = \frac{AC}{AI_a}. \quad ②$$

From ① and ② we get

$$\frac{AF}{AI} = \frac{AI_a}{AE}. \quad ③$$

By the assumption $\angle FAI_a = \angle EAI$, we have $\triangle AFI_a \backsim \triangle AIE$, and therefore $\angle FI_aA = \angle IEA$, which completes the proof.

3 Let $\mathbf{N}$ be the set of positive integers. Determine all functions $g: \mathbf{N} \to \mathbf{N}$ such that $(g(m) + n)(m + g(n))$ is a perfect square for all $m, n \in \mathbf{N}$. (Posed by America)

Solution The answer is $g(n) = n + C$, where C is a nonnegative integer.

Clearly, the function $g(n) = n + C$ satisfies the required property since

$$(g(m) + n)(m + g(n)) = (n + m + C)^2$$

is a perfect square. We first prove a lemma.

Lemma If a prime number p divides $g(k) - g(l)$ for some positive integers k and l, then $p \mid k - l$.

Proof of lemma: If $p^2 \mid g(k) - g(l)$, let $g(l) = g(k) + p^2 a$, where a is an integer. Choose an integer

$$D > \max\{g(k), g(l)\},$$

and D is not divisible by p. Set $n = pD - g(k)$; then $n + g(k) = pD$, and thus

$$n + g(l) = pD + (g(l) - g(k)) = p(D + pa)$$

is divisible by p, but not divisible by p^2.

By assumption, $(g(k) + n)(g(n) + k)$ and $(g(l) + n)(g(n) + l)$ are both perfect squares, and therefore they are divisible by p^2 since they are divisible by p. Hence,

$$p \mid ((g(n) + k) - (g(n) + l)),$$

i.e. $p \mid k - l$.

If $p \mid g(k) - g(l)$ but p^2 does not divide $g(k) - g(l)$, choose an integer D as above and set $n = p^3 D - g(k)$. Then $g(k) + n = p^3 D$ is divisible by p^3, but not by p^4, and

$$g(l) + n = p^3 D + (g(l) - g(k))$$

is divisible by p, but not by p^2. As with the above argument, we have $p \mid g(n) + k$ and $p \mid g(n) + l$, and therefore

$$p \mid ((g(n)+k)-(g(n)+l)),$$

i.e. $p \mid k-l$. This completes the proof of the lemma.

Back to the original problem: if there exist positive integers k and l such that $g(k) = g(l)$, then the lemma implies that $k-l$ is divisible by any prime number. Hence, $k-l=0$, i.e. $k=l$, and thus g is injective.

Now consider $g(k)$ and $g(k+1)$. Since $(k+1)-k=1$, once again the lemma implies that $g(k+1)-g(k)$ is not divisible by any prime number, and therefore

$$|g(k+1)-g(k)|=1.$$

Let $g(2)-g(1)=q$, where $|q|=1$. It follows easily by induction that

$$g(n) = g(1)+(n-1)q.$$

If $q=-1$, then $g(n) \leqslant 0$ for $n \geqslant g(1)+1$, a contradiction. Therefore, we must have $q=1$ and

$$g(n) = n+(g(1)-1)$$

for any $n \in \mathbf{N}$, where $g(1)-1 \geqslant 0$. Set $g(1)-1=C$ (a constant). Then $g(n) = n+C$, where C is a nonnegative integer.

Second Day

(0900 - 1330; July 8, 2010)

4 Let P be a point inside the triangle ABC. The lines AP, BP and CP intersect the circumcircle Γ of the triangle ABC again at the points K, L and M respectively. The tangent to Γ at C intersects the line AB at S. Suppose that $SC = SP$. Prove that $MK = ML$. (Posed by Poland)

Proof Without loss of generality, we assume that $CA > CB$. Then S lies on the extension of AB. Suppose that the line SP intersects the circumcircle of the triangle ABC at E, F, as in the figure. By assumption and the power of a point theorem, we have

$$SP^2 = SC^2 = SB \times SA,$$

and hence $\frac{SP}{SB} = \frac{SA}{SP}$. Then $\triangle PSA \backsim \triangle BSP$ and $\angle BPS = \angle SAP$.

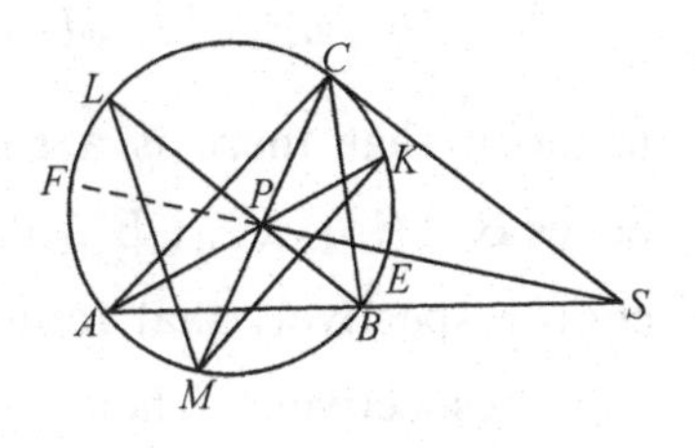

Since $2\angle BPS = \overset{\frown}{BE} + \overset{\frown}{LF}$, $2\angle SAP = \overset{\frown}{BE} + \overset{\frown}{EK}$, we have

$$\overset{\frown}{LF} = \overset{\frown}{EK}. \quad ①$$

It follows from $\angle SPC = \angle SCP$ that $\overset{\frown}{EC} + \overset{\frown}{MF} = \overset{\frown}{EC} + \overset{\frown}{EM}$, and therefore we have

$$\overset{\frown}{MF} = \overset{\frown}{EM}. \quad ②$$

From ① and ②, we get

$$\overset{\frown}{MFL} = \overset{\frown}{MF} + \overset{\frown}{FL} = \overset{\frown}{ME} + \overset{\frown}{EK} = \overset{\frown}{MEK},$$

and thus $MK = ML$.

5 In each of the six boxes B_1, B_2, B_3, B_4, B_5, B_6, there is initially one coin. There are two types of operation allowed:

Type 1: Choose a nonempty box B_j, with $1 \leqslant j \leqslant 5$. Remove one coin from B_j and add two coins to B_{j+1}.

Type 2: Choose a nonempty box B_k, with $1 \leqslant k \leqslant 4$. Remove one coin from B_k and exchange the contents of the (possibly empty) boxes B_{k+1} and B_{k+2}.

Determine whether there is a finite sequence of such operations that results in the boxes B_1, B_2, B_3, B_4, B_5 being empty and the box B_6 containing exactly $2010^{2010^{2010}}$ coins. (Note that $a^{b^c} = a^{(b^c)}$.) (Posed by Holland)

Solution The answer is affirmative.

Let $A = 2010^{2010^{2010}}$. Denote by

$$(b_i, b_{i+1}, \ldots, b_{i+k}) \to (b'_i, b'_{i+1}, \ldots, b'_{i+k})$$

to mean that there is a finite sequence of operations on the boxes B_i, B_{i+1}, ..., B_{i+k} initially containing b_i, b_{i+1}, ..., b_{i+k} coins respectively that results in containing b'_i, b'_{i+1}, ..., b'_{i+k} coins respectively. Then we shall show that

$$(1, 1, 1, 1, 1, 1) \to (0, 0, 0, 0, 0, A).$$

Lemma 1 *For every positive integer* a, *we have* $(a, 0, 0) \to (0, 2^a, 0)$.

Proof of lemma 1: We prove by induction on $k \leqslant a$ that $(a, 0, 0) \to (a-k, 2^k, 0)$.

Since $(a, 0, 0) \to (a-1, 2, 0) \to (a-1, 2^1, 0)$, the assertion is true for $k = 1$. Suppose that the assertion is true for some $k < a$; then

$$(a-k, 2^k, 0) \to (a-k, 2^k-1, 2) \to \cdots \to (a-k, 0, 2^{k+1}) \to (a-k-1, 2^{k+1}, 0),$$

and thus

$$(a, 0, 0) \to (a-k, 2^k, 0) \to (a-k-1, 2^{k+1}, 0).$$

The assertion is also true for $k+1 (\leqslant a)$. By the method of induction. Lemma 1 is proven.

Lemma 2 *For every positive integer* a, *we have* $(a, 0,$

0, 0)$\to$(0, P_a, 0, 0), *where* $P_n = 2^{2^{\cdot^{\cdot^{2}}}}$ (*with* n 2*'s*) *for a positive integer* n.

Proof of Lemma 2: We prove by induction on $k \leqslant a$ that $(a, 0, 0, 0) \to (a-k, P_k, 0, 0)$.

By the operation of type 1, we have

$$(a, 0, 0, 0) \to (a-2, 2, 0, 0) = (a-1, P_1, 0, 0),$$

and the assertion is true for $k = 1$. Suppose that the assertion is true for some $k < a$; then

$$\begin{aligned}&(a-k, P_k, 0, 0) \to (a-k, 0, 2^{P_k}, 0)\\ &= (a-k, 0, P_{k+1}, 0) \to (a-k-1, P_{k+1}, 0, 0),\end{aligned}$$

and therefore

$$(a, 0, 0, 0) \to (a-k, P_k, 0, 0) \to (a-k-1, P_{k+1}, 0, 0),$$

i.e. the assertion is also true for $k+1 \leqslant a$. By the method of induction, we have proven Lemma 2.

We have

$$\begin{aligned}&(1, 1, 1, 1, 1, 1) \to (1, 1, 1, 1, 0, 3) \to (1, 1, 1, 0, 3, 0)\\ &\to (1, 1, 0, 3, 0, 0) \to (1, 0, 3, 0, 0, 0) \to (0, 3, 0, 0, 0, 0) \to\\ &(0, 0, P_3, 0, 0, 0) = (0, 0, 16, 0, 0, 0) \to (0, 0, 0, P_{16}, 0, 0),\end{aligned}$$

and

$$\begin{aligned}A &= 2010^{2010^{2010}} < (2^{11})^{2010^{2010}}\\ &= 2^{11\times 2010^{2010}} < 2^{2010^{2011}} < 2^{(2^{11})^{2011}}\\ &= 2^{2^{11\times 2011}} < 2^{2^{2^{15}}} < P_{16},\end{aligned}$$

and thus the number of coins in the box B_4 is greater than A. Therefore, by performing operations of type 2, we have

$$\begin{aligned}&(0, 0, 0, P_{16}, 0, 0) \to (0, 0, 0, P_{16}-1, 0, 0)\\ &\to (0, 0, 0, P_{16}-2, 0, 0)\end{aligned}$$

$$\to \cdots \to \left(0, 0, 0, \frac{A}{4}, 0, 0\right).$$

Then we get

$$\left(0, 0, 0, \frac{A}{4}, 0, 0\right) \to \left(0, 0, 0, 0, \frac{A}{2}, 0\right) \to (0, 0, 0, 0, 0, A).$$

6 Let $a_1, a_2, a_3, \ldots$ be a sequence of positive real numbers. Suppose that for some positive integer s, we have

$$a_n = \max\{a_k + a_{n-k} \mid 1 \leqslant k \leqslant n-1\} \qquad ①$$

for all $n > s$. Prove that there exist positive integers l and N, with $l \leqslant s$ and such that $a_n = a_l + a_{n-l}$ for all $n \geqslant N$. (Posed by Iran)

Proof By assumption, for each $n > s$, a_n can be written as $a_n = a_{j_1} + a_{j_2}$, $j_1, j_2 < n$, $j_1 + j_2 = n$. If $j_1 > s$, we can continue to write a_{j_1} as a sum of two terms of the sequence. We keep doing this until we obtain

$$a_n = a_{i_1} + \cdots + a_{i_k}, \qquad ②$$

$$1 \leqslant i_j \leqslant s,\ i_1 + \cdots + i_k = n. \qquad ③$$

Suppose that a_{i_1}, a_{i_2} is the last step in obtaining ②. Then $i_1 + i_2 > s$, and ③ is reformulated as

$$1 \leqslant i_j \leqslant s,\ i_1 + \cdots + i_k = n,\ i_1 + i_2 > s. \qquad ④$$

On the other hand, if the indices $i_1, \ldots, i_k$ satisfy ④, we set $s_j = i_1 + \cdots + i_j$. By ①, we have

$$a_n = a_{s_k} \geqslant a_{s_{k-1}} + a_{i_k} \geqslant a_{s_{k-2}} + a_{i_{k-1}} + a_{i_k} \geqslant \cdots \geqslant a_{i_1} + \cdots + a_{i_k}.$$

Hence, for any $n > s$, we get

$$a_n = \max\{a_{i_1} + \cdots + a_{i_k} : (i_1, \ldots, i_k) \text{ satisfies ④}\}.$$

Let $m = \max\left\{\frac{a_i}{i} \mid 1 \leqslant i \leqslant s\right\}$, and assume that $m = \frac{a_l}{l}$ for some positive integer $l \leqslant s$.

Construct a sequence $\{b_n\}$ as follows: $b_n = a_n - mn$, $n = 1, 2, \ldots$; then $b_l = 0$. When $n \leqslant s$, we have $b_n \leqslant 0$ by definition of m. When $n > s$,

$$\begin{aligned} b_n &= a_n - mn = \max\{a_k + a_{n-k} \mid 1 \leqslant k \leqslant n-1\} - mn \\ &= \max\{b_k + b_{n-k} + mn \mid 1 \leqslant k \leqslant n-1\} - mn \\ &= \max\{b_k + b_{n-k} \mid 1 \leqslant k \leqslant n-1\} \leqslant 0, \end{aligned}$$

so $b_n \leqslant 0$, $n = 1, 2, \ldots$, and for $n > s$,

$$b_n = \max\{b_k + b_{n-k} \mid 1 \leqslant k \leqslant n-1\}.$$

If $b_k = 0$ for every $k = 1, 2, \ldots, s$, then for every positive integer n, $b_n = 0$, and hence $a_n = nm$ for every n, and the conclusion follows.

Otherwise, let $M = \max\limits_{1 \leqslant i \leqslant s} |b_i|$, $\varepsilon = \min\{|b_i| : 1 \leqslant i \leqslant s, b_i < 0\}$. When $n > s$, we have

$$b_n = \max\{b_k + b_{n-k} \mid 1 \leqslant k \leqslant n-1\} \geqslant b_l + b_{n-l} = b_{n-l},$$

and thus $0 \geqslant b_n \geqslant b_{n-l} \geqslant \cdots \geqslant -M$.

As for the sequence $\{b_n\}$, by ② and ③ every b_n belongs to the set

$$T = \{b_{i_1} + b_{i_2} + \cdots + b_{i_k} : 1 \leqslant i_1, \ldots, i_k \leqslant s\} \cap [-M, 0].$$

It follows that T is a finite set. Indeed, for any $x \in T$, let

$$x = b_{i_1} + \cdots + b_{i_k} \,(1 \leqslant i_1, \ldots, i_k \leqslant s).$$

Then there are at most $\frac{M}{\varepsilon}$ nonzero terms in b_{i_j} [otherwise $x <$

$\frac{M}{\varepsilon}(-\varepsilon) = -M]$, and hence there are only finitely many such representations for x.

Thus, for every $t = 1, 2, \ldots, l$, the sequence

$$b_{s+t}, b_{s+t+l}, b_{s+t+2l}, \ldots$$

is increasing, and takes only finitely many values, and hence it is constant eventually, i.e. there is N such that $\{b_n\}$ is periodic for $n > N$ with period l; in other words,

$$b_n = b_{n-l} = b_l + b_{n-l} (n > N + l),$$

i.e.

$$\begin{aligned} a_n &= b_n + mn = (b_l + ml) + (b_{n-l} + m(n-l)) \\ &= a_l + a_{n-l} (n > N + l). \end{aligned}$$